U0156086

景德镇

JING DE ZHEN

起于汉唐
Starting in the Han and Tang Dynasties

成于宋元
Maturing in the Song and Yuan Dynasties

盛于明清
Attaining Its Acme in the Ming and Qing Dynasties

兴于当代
Reviving in the Contemporary Era

景德镇瓷业发展简史

A Brief History of Jingdezhen Porcelain Industry

魏望来　王丽心　著

朱练平　译

Writer: Wei Wanglai and Wang Lixin

Translator: Zhu Lianping

中国文史出版社

CHINA CULTURAL AND HISTORICAL PRESS

目 录
Contents

第二编　成于宋元

II　Maturing in the Song and Yuan Dynasties

第三编　盛于明清

III　Attaining Its Acme in the Ming and Qing Dynasties

第四编　兴于当代
IV　Reviving in the Contemporary Era

第一编

起于汉唐

I

Starting in the Han and Tang Dynasties

水土宜陶

A Promised Land for Ceramic Manufacturing

景德镇因瓷而生，因瓷而名，因瓷而兴。

从漫漫远古，景德镇起航。

Jingdezhen was born because of porcelain. It is famous and prosperous because of porcelain. Since ancient times, Jingdezhen has started to produce porcelain.

景德镇原始森林

Primary Forest in Jingdezhen City

高岭矿坑
Kaolin Mining Pit

景德镇市位于江西省东北部，下辖乐平市、浮梁县、珠山区、昌江区，市域范围东经 116°57′~117°42′，北纬 28°42′~29°56′。北部与安徽省池州市和黄山市交界，东、南、西三面与江西省上饶市毗邻。土地面积为 5261 平方千米，森林覆盖率为 67.85%，2019 年末常住人口为168.05 万人。

Jingdezhen is situated in the northeast of Jiangxi Province and borders on Chizhou City and Huangshan City in Anhui Province in the north, and Shangrao City in Jiangxi Province in the east, south and west; The territory of Jingdezhen lies between longitudes 116°57′ to 117°42′ East, and latitudes 28°42′ to 29°56′ North. Its area is 5,261 square kilometers with the forest coverage rate reaching 67.85% and a permanent resident population of 168.05 million at the end of 2019. Leping City, Fuliang County, Changjiang District and Zhushan District are under its jurisdiction.

距今 10 亿~14 亿年前，景德镇还浸没于一片汪洋大海之中。距今 1 亿~5 亿年前，景德镇逐渐从海洋中隆起，演绎了沧海桑田巨变。地质运动造就景德镇市域东北高、西南低、形如筲箕的中低丘陵地貌，形成状如叶脉的山川河流，积聚起丰富的瓷土矿藏，构成了陶瓷生产必需的全部自然资源供给系统。

1-1.4 billion years ago, Jingdezhen was still immersed in a vast ocean. 100-500 million years ago, Jingdezhen was gradually uplifted from the ocean, demonstrating great changes. The terrain of Jingdezhen is high in the southwest and low in the northeast with low hilly landform shaped like Shaoji（a bamboo basin kitchen utensil for washing rice or vegetable）which was created by the geological movement, forming leaf-vein shape mountains and rivers and accumulating rich porcelain clay minerals which constitutes the supply system of all natural resources necessary for ceramic production.

五行相加而生陶焉。冥冥之中，似有天佑。

Heaven has produced the five elements-wood, fire, earth, metal, and water-are believed to be the primary components of porcelain-not altogether incidentally, all of which are easily available in the city.

景德镇属亚热带季风气候，雨量充沛，日照充足，无霜期长，四季分明，对自然植被生长、水路运输、水力利用和瓷业生产十分有利。景德镇年平均气温为 17.1℃，温差变化小，瓷器坯胎不易胀缩。年平均无霜期为 248 天，年平均降水量为 1763.5 毫米，总量适中。年平均日照时数为 1968.5 小时，瓷器坯胎不易潮湿变形，也容易晾干。

Jingdezhen has a subtropical monsoon climate with abundant rainfall, sufficient sunshine, long frost-free period and four distinct seasons, which is very beneficial to the growth of natural vegetation, waterway transportation, hydraulic utilization and porcelain production. With an annual average temperature 17.1℃, clay body is not easy to expand and shrink because of little change of temperature difference. The average frost-free period over the years is 248 days, the average annual precipitation is 1,763.5 mm, and the total amount is moderate. The city receives 1,968.5 hours of bright sunshine annually. In this climate, the clay body is not easy to be deformed by being exposed to moisture, but also easy to dry.

自然环境是古代陶瓷生产的先决条件。

Natural environment is a prerequisite for the production of ancient ceramics.

经历多次地质构造运动，景德镇市域形成了不同时代的地层、构造、岩浆岩以及相关矿产资源。岩体经过风化、蚀变，形成了风化残积型高岭土和脉状高岭土矿床。除了花岗岩体，还有长英岩、微晶花岗岩、细晶岩、长石石英斑岩、石英长石斑岩等脉状岩浆岩。岩体风化蚀变后，形成软质或硬质瓷石矿床，是陶瓷生产的主要矿产资源。景德镇市域蕴藏的瓷业原料资源品种达 40 多种。

Having experienced many geological tectonic movements, strata, structures, magmatic rocks and related mineral resources of different ages have been formed

码堆柴垛

Stacked Kiln Firewood

in Jingdezhen city. With weathering and alteration, the rock mass forms weathered residual kaolin and vein kaolin deposits. Besides granite body, there are vein mag matic rocks such as felsic, microcrystalline granite, fine-grained rock, feldspar quartz porphyry and quartz feldspar porphyry. After weathering and alteration of rock mass, soft or hard porcelain stone deposits are formed, which are the main mineral resources for ceramic production. There are more than 40 varieties of porcelain raw material resources in Jingdezhen city.

景德镇为竹木之乡，木本植物有 95 科 350 多种，马尾松是其中的优势种群，松木锯成的烧柴是瓷窑的理想燃料，温度可达 1300℃。古代景德镇陶瓷生产就地取材，节省运费，瓷器品质好，产量高，提高了竞争力。

Jingdezhen is a home of bamboo and wood abound with more than 350 woody plants in 95 families, among which the dominant species is masson pine. The firewood sawn from pine trees is an ideal fuel for porcelain kilns with firing temperature reaching 1,300℃. In ancient times, the production of ceramics from local raw materials in Jingdezhen could save freight cost. Moreover, the local porcelain products are characterized with good quality and high yield, which improved its competitiveness.

水系发达是古代陶瓷生产的必备条件。

Well-developed water system is an essential prerequisite for the production of ancient ceramics.

陶瓷生产中的瓷石粉碎、澄清瓷土、泥料淘洗、制造瓷胎、调配色釉等环节，均需大量用水。景德镇市域水系分布均匀，水量充沛，含沙量小，杂质少，可确保瓷器质量。主流昌江贯穿全境，河流在景德镇境内全长 115 公里，支流东河、南河、西河、小北河、梅湖河、建溪河等 50 多条大

水 碓

A Water-powered Trip Mill

小河流汇集于昌江，构成通达各个乡村的水上运输网络。上游的瓷土、窑柴和木炭可顺流而下，直抵景德镇。

A large amount of water is needed in the processes of porcelain stone crushing, clarification of porcelain clay, clay washing, the fabrication of biscuits, the preparation of color glaze and so on. Jingdezhen water system is evenly distributed with abundant water, small sand content and less impurities, which can ensure the quality of porcelain wares. The mainstream of the Chang River runs through the whole territory with a flow length of 115 kilometers in Jingdezhen. Its tributaries, more than 50 large and small rivers, including the East River, the South River, the West River, the Xiaobei River, the Meihu River and the Jianxi River, converge and flow in the Chang River, forming a water transportation network connecting all the villages. The porcelain clay, kiln firewood and charcoal in the upstream of the river float downward with the current until it reaches to Jingdezhen.

缪宗周《兀然亭》说："陶舍重重倚岸开，舟帆日日蔽江来。"

Miu Zongzhou (an official of the Ming Dynasty) wrote in a poem titled "Wuran Pavilion" that "Kilns are arranged along the rivers, and boats and ships which carry porcelain wares come and go every day."

景德镇昌江之畔
The Riverside of the Chang River in Jingdezhen

　　没有城墙的景德镇，处于碧水涟漪的昌江两岸，南河环绕于东南，西河贯穿于西岸，形成了三水环城之势。山环水绕，景色秀丽，古老的景德镇宛如一颗明珠镶嵌其中。先民沿河建窑，沿窑成市。制成的瓷器装船扬帆，顺着昌江，经鄱阳湖，分流到各大水系，直至通江达海，远涉外洋。

　　Without encircling wall, Jingdezhen is located on both sides of the Chang River with clear water ripples. The South River surrounds the southeast, and the West River runs through the west bank, forming a trend of three rivers around the city. The ancestors here built kilns along the river and set up markets along the kilns. After loaded on ships, the finished porcelains were transported along the Chang River, and then diverted to major water systems through Lake Poyang until it traveled across vast oceans.

开放格局是集约化陶瓷生产的地理条件。

Opening-up is the geographical condition of intensive ceramic production.

所以，《景德镇陶录》记载："水土宜陶，陈以来土人多业于此。"

Therefore, it is recorded in *Records of Jingdezhen Ceramics* that "Clay and water are suitable for ceramics, and the natives have worked here since the Chen Dynasty."

始于汉世

Porcelain Manufacturing Traced Back to the Han Dynasty

浮梁古城
The Old City of Fuliang

历史上，景德镇长期归属浮梁县管辖。

Historically, Jingdezhen had long been under the jurisdiction of Fuliang County.

春秋战国时期，浮梁地属古番，春秋时为楚东南境。东周姬匄之敬王十六年（前504年），吴伐楚取番。越王勾践打败吴王夫差后，属越。楚灭越后，浮梁境复归楚，此即浮梁"吴头楚尾"之说。秦嬴政之始皇帝二十六年（前221年），浮梁地属九江郡番县。西汉刘邦之高帝五年（前202年），番县改称番阳县，属豫章郡。东汉刘秀之光武帝时（25～57年），番阳县改称鄱阳县，属庐陵郡。

During the Spring and Autumn Period and the Warring States Period, Fuliang was under the jurisdiction of the ancient Fan, and it was the southeast of the Chu State during the Spring and Autumn period. In the sixteenth year of King Jing (personal name Jigai) of the Eastern Zhou Dynasty (504 B.C.), the Wu State conquered the Chu State and took Fan. After Gou Jian, king of the Yue State, defeated Fu Chai, king of the Wu State who belonged to the Yue State. After the Chu State destroyed the Yue State, the territory of Fuliang returned to the Chu State; hence it was called "the head of the Wu Kingdom and the tail of the Chu Kingdom." In the twenty-sixth year of the first emperor (personal name Yingzheng) of Qin Dynasty (221 B.C.), Fuliang belonged to Fan County, Jiujiang Commandery. In the fifth year of Emperor Gaodi (personal name Liu Bang) of the Western Han Dynasty (202 B.C.), Pan County was renamed as Fanyang County, belonging to Yuzhang Commandery. During the period of Emperor Guangwu (personal name Liu Xiu) of the Eastern Han Dynasty (25-57), Fanyang County was renamed as Poyang County, belonging to Luling Commandery.

景德镇市域存有先民早期制陶遗迹。

There are relics of early pottery making by ancestors in Jingdezhen city.

在浮梁县王港乡水家村水家车文化遗址，发现有陶器等遗存，鉴定为新石器时期遗址。在蛟潭镇，一处2000平方米的商周时期文化遗址中，出土瓮、樽、折肩罐、杯盏等陶器，质地为红褐色夹砂陶、软质泥灰陶、硬质泥灰陶三种。在湘湖镇东流村，发现 1.5 万平方米的商周时期原始瓷遗址，遗物有陶器和原始瓷。

Pottery and other relics were found at the Shuijiache Cultural Site in Shuijia village, Wanggang township, Fuliang County, which was identified as a Neolithic site. At Jiaotan town, in a 2000-square-metercultural site of the Shang and Zhou dynasties, pottery wares, including urn, zun, jar with angular shoulder and wine cups, were unearthed, the texture of which is classified into three kinds, including reddish brown sand mixed pottery, soft and hard marl pottery. At Dongliu village, Xianghu town, a primitive porcelain relic site of 15,000 square meters of the Shang and Zhou dynasties were found with relics including pottery and primitive porcelain.

汉代至隋代景德镇制瓷情况，仅见于史籍记载。

The manufacturing of porcelain in Jingdezhen from the Han Dynasty to the Sui Dynasty can only be found in historical records.

浮梁昌江两岸
Both Sides of the Chang River in Fuliang County

《浮梁县志》记载："新平治陶，始于汉世，大抵坚重朴茂，范土合埏，有古先遗制。"《南窑笔记》亦载："新平之景德镇，在昌江之南，其治陶始于季汉。"汉代分西汉（前 206～25 年）和东汉（25～220 年）。季汉，当指东汉。

According to *Fuliang County Chronicle*, Jingdezhen, "formerly Xinping, began to produce porcelain during the Han Dynasty. Generally, Xinping pottery wares were characterized by heavy shape and dense body, which was made according to the method inherited from the previous generation." As also recorded in *Nan Kiln Notes*, "Situated on the south bank of the Chang River, Jingdezhen, originally called Xinping, started its ceramic production in Ji Han." The Han dynasty is divided into two phases: the Western Han (206 B.C.-A.D.25) and the Eastern Han (25-220). Ji Han is the so-called the Eastern Han Dynasty.

《浮梁县志》载："东晋于昌南设新平镇。"又载："陶侃擒江东寇于昌南，遂改昌南为新平镇。"时间当在东晋成帝司马衍之咸和五年（330年）。新平镇并非官方认可的行政建制，只是军事意义上的镇守。到了唐玄宗李隆基之开元四年（716 年），浮梁县制恢复并更名新昌县后，因县治迁至新昌江口之南城（今浮梁镇新平村），而新平镇在县治新昌江口以南地域，所以文人又雅称新平镇为"昌南镇"。

As noted in *Fuliang County Chronicles*, "Xinping town was set up at Changnan in the Eastern Jin Dynasty"; "Tao Kan defeated the rebels from Jiangdong area at Changnan, and then changed Changnan to Xinping town", which was in the fifth year of the Xianhe reign of Emperor Chengdi (personal name Sima Yan) of the Eastern Jin Dynasty (330). Xinping town was not an officially recognized administrative region, but a garrison in the military sense. Until the fourth year of the Kaiyuan reign of Emperor Xuanzong (personal name Li Longji) of the Tang Dynasty (716), the county system of Fuliang was restored, which was renamed as Xinchang, and the county government was moved to Nancheng (now Xinping village, Fuliang town) in the mouth of the Xinchang River. Because Xinping town is located in the south of the Xinchang River mouth, it is also called Changnan town by scholars.

如果从东汉初年算起，景德镇有 2000 年制瓷史。

If calculated from the early years of the Eastern Han Dynasty, Jingdezhen has a porcelain making history of 2,000 years.

汉代景德镇制造的当属原始瓷器或早期瓷器，器物粗糙厚实，釉色偏黄和微黑。这类早期瓷器瓷质不纯，质量也不高，还处于"耕而陶"的发展阶段，即制瓷手工业与农业结合在一起，产品也主要是卖给邻近的乡民，并不远销。

These artifacts made in Jingdezhen in the Han Dynasty are considered as proto-porcelains or early porcelains, which are marked by rough and thick body with yellowish and slightly black glaze. The early porcelain was impure in texture, and its quality was of very low, which was still in the development stage of "making pottery after farming". In another word, handicraft industry of porcelain manufacturing was combined with agriculture, and the products were mainly sold to nearby villagers and were not sold far away.

三国时，鄱阳郡属吴国扬州，浮梁地域随之隶属扬州管辖。西晋惠帝司马衷之元康元年（291 年），鄱阳郡改属江州。南北朝之梁元帝萧绎之承圣二年（553 年），浮梁随鄱阳郡复属江州。

During the Three Kingdoms Period, Poyang Commandery belonged to Yangzhou of the Wu State, so Fuliang region was under the jurisdiction of Yangzhou. In the first year of the Yuankang reign of Emperor Huidi (personal name Sima Zhong) of the Eastern Jin Dynasty (291), Poyang Commandery was subordinate to Jiangzhou. In the second year of the Chensheng reign of Emperor

Yuandi (personal name Xiao Yi) of the Liang Dynasty during the Southern and Northern Dynasties period (553), Fuliang thereafter returned to Jiangzhou with Poyang Commandery.

三国、两晋、南北朝时期，景德镇治陶跃升至瓷器，耕陶开始分离。

During the Three Kingdoms, the Two Jins and the Southern and Northern Dynasties, Jingdezhen started to make porcelain instead of primitive porcelain. Meanwhile, porcelain making was separating from farming.

东晋时，有位名叫赵慨的人隐居于新平镇。他运用在浙江为官时掌握的越窑制瓷技艺，对当地制瓷的胎釉配制、成型、焙烧等工艺进行一系列重大改革，为瓷业发展作出重要贡献，先民尊他为制瓷师主，后人建师主庙奉祀。

During the Eastern Jin Dynasty, a man named Zhao Kai lived in seclusion in Xinping town. By using skills that he had mastered when he worked as an official in Zhejiang, he carried out a series of major reforms in techniques of local porcelain making, including clay body and glaze preparation, shaping, firing and other processes of local porcelain making, which made important contributions to the development of porcelain manufacturing. The ancient ancestors respected him as "the master of porcelain making", and later generations built a master temple to worship him.

浮梁东埠
Dongbu, Fuliang County

景德镇生产贡瓷的最早记载在陈朝。

Porcelain wares made in Jingdezhen specially for use at court were first recorded in the Chen Dynasty.

《景德镇陶录》记载："陈至德元年，诏镇以陶础贡建康。""镇陶自陈以来名天下。"说的是南北朝之陈后主之至德元年（583 年），陈叔宝扩建皇宫，下诏新平镇制作陶瓷柱础上贡建康（今江苏省南京市）。陶瓷柱础因强度达不到要求，承受不了高大阁柱的沉重压力，虽然巧而弗坚，不堪用。但新平镇的名声却由此传播开来。

As recorded in *Records of Jingdezhen Ceramics*, "In the first year of the Zhide reign of the Chen Dynasty, the town was ordered to build pottery pillars as tributes to Jiankang." "Jingdezhen porcelain has been famous all over the world since the Chen dynasty." It means that in the first year of the Zhide reign of the Chen Dynasty-the last dynasty of the Southern and Northern Dynasties (583), Chen Shubao, often known in history as Houzhu of Chen (literally "Chen's final lord"), issued an edict that Xinping town supplied ceramic pillars as tributes to Jiankang (now Nanjing, Jiangsu Province) for the extension of the imperial palace. These tributary ceramic pillars from the town could not bear the heavy pressure of huge pavilion columns, for the strength could not meet the requirements, although it had the characteristics of light and handy volume. Since then, the reputation of Xinping town had spread near and far.

《南窑笔记》记载，隋炀帝杨广之大业年间（605～617 年），奉朝廷之命，新平镇瓷工制成两座瓷质狮象大兽，送往了洛阳，贡奉于金碧辉煌的显仁宫。

According to the *Nan Kiln Notes*, during the Daye reign of the Sui Dynasty (605-617), potters in Xinping town fashioned two statuary porcelains of lion and elephant by imperial order, which were sent to Luoyang and enshrined in the magnificent Xianren Palace.

制器进御

Porcelain as Tribute to the Court

唐高祖李渊之武德二年（619 年），于鄱阳东界置新平乡。武德四年（621 年），从鄱阳县分析设置新平县。武德八年（625 年），县制减并，重入鄱阳县。唐玄宗李隆基之开元四年（716 年），恢复县制并更县名为新昌。天宝元年（742 年），更新昌县为浮梁县，地域属鄱阳郡。

In the second year of the Wude reign of Emperor Gaozu (personal name Li Yuan) of the Tang Dynasty (619), Xinping township was set up in the east boundary of Poyang. In the fourth year of the Wude reign (621), Xinping County was set up by separating from Poyang County, which was incorporated into as a part of Poyang County because of the subtraction and merger of the county system in the eighth year of the Wude reign (625). In the second year of the Kaiyuan reign of Emperor Xuanzong (personal name Li Longji) of the Tang Dynasty (716), the county system was restored and the county was renamed as Xinchang. In the first year of the Tianbao reign (742), Xinchang was renewed as Fuliang County, belonging to Poyang Commandery.

古城余晖
Afterglow of the Old City

考古发掘表明，景德镇瓷业始于中晚唐。

Archaeological excavation shows that Jingdezhen porcelain manufacturing started in the middle and late Tang Dynasty.

青瓷是唐代景德镇窑的主要产品。此时，新平镇瓷工掌握了高火度烧瓷方法，所制瓷器色泽素润，质地坚固，出现了一些烧造瓷器的名窑。史料记载，镇民陶玉的陶窑所产瓷器"土惟白壤，体稍薄，色素润"。陶玉将所产瓷器载入关中，进贡于朝，被称为"假玉器"。镇民霍仲初的霍窑所产瓷器"瓷色亦素，土壤腻，质薄，佳者莹缜如玉"。不过，陶窑和霍窑遗址均未发现。

Celadon was the main product of Jingdezhen kilns in the Tang Dynasty. Since then, potters in Xinping town had already mastered the skill of firing porcelain at a much higher temperature. The kind of porcelain was characterized by plain in color and solid in texture. During this period, some famous kilns for firing porcelain emerged. According to the historical records, porcelain wares from "Tao Kiln" (named after Tao Yu) and "Huo Kiln" (named after Huo Zhongchu) were called "artificial jade", which were sent to the imperial court as tribute because of hard body, fine texture, and jade-like glaze. The porcelain objects made by Tao Kiln and Huo Kiln should be categorized as plain porcelain, namely celadon. However, none of Tao Kiln and Huo Kiln sites have been uncovered.

唐武德四年（621年），高祖李渊下达诏令，命新平霍仲初等"制瓷进御"。朝廷还在新平镇设立监务所，以督理陶务。

In the fourth year of the Wude reign of the Tang Dynasty (621), Emperor Gaozu (personal name Li Yuan) issued an imperial edict ordering Huo Zhongchu together with other potters to "make porcelain as tribute to the court". The court also established a supervision office in Xinping town to oversee ceramic affairs.

乐平南窑龙窑遗迹
Dragon Kiln Ruins at Nanyao, Leping City

乐平南窑唐代遗址发现于 1964 年，是景德镇市域最早的瓷业遗存，始烧于中唐，由此把景德镇制瓷史推进了 200 年。南窑遗址是我国迄今发现窑炉分布最密集、布局最有规律、瓷业组织最严密的唐代窑场，也是最长的唐代龙窑遗迹，填补了景德镇瓷器烧造窑炉形制最早形态的空白。

Nanyao kiln site, located in Leping, was discovered in 1964. Started firing in the middle Tang Dynasty, the site was proved to be the earliest ceramic industry site discovered in Jingdezhen and pushed the history of porcelain manufacturing in Jingdezhen back 200 years. Nanyao kiln site is the kiln remains site of the Tang Dynasty with the densest distribution of kilns, the most regular layout and the most rigorous porcelain industry organization found so far in China. It is also the longest Tang Dynasty dragon kiln site, filling the gap in the earliest form of kilns for firing porcelain in Jingdezhen.

唐肃宗李亨之乾元二年（759 年），饶州刺史颜真卿巡行新平，与陆士修等名士月夜品茗，合作《五言月夜啜茶联句》："泛花邀坐客，代饮引情言。醒酒宜华席，留僧想独园。不须攀月桂，何假树庭萱。御史秋风劲，尚书北斗尊。流华净肌骨，疏瀹涤心原。不似春醪醉，何辞绿菽繁。素瓷传静夜，芳气满闲轩。"

In the second year of the Qianyuan reign of Emperor Suzong (personal name Li Heng) of the Tang Dynasty (759), when Yuan Zhengqing, the governor of Raozhou, inspected Xinping, where he had tea with some famous scholars including Lu Shixiu and other friends on a moon night, they composed a joint authorship poem titled "Sipping tea at a moonlit night in five-character verse": "Friends are invited enjoy flowers and exchange feelings at home, using tea as a substitute for wine. A drunken man tends to miss the epicurean feast when waking up while a monk left behind only think of the quiet and lonely temple. Only those who don't pursue fame, money and material goods can be carefree. Being an imperial censor, Yan Zhenqing's integrity is firm and admirable. Drinking tea under the bright moonlight makes people feel relaxed, refined and detached from reality. Although there is no intoxicating wine of the imperial court among the populace, ordinary people can also take much delight in farming and working. The tea fragrance of the white porcelain cup slowly diffuses into the quiet night and permeates the whole courtyard, creating a sense of calm and wellbeing."

唐宪宗李纯之元和八年（813 年），饶州刺史元崔督造瓷器向朝廷进贡，请柳宗元代作了《进瓷器状》："瓷器若干事。右件瓷器等，并艺精埏埴，制合规模。禀至德之陶蒸，自无苦窳；合太和以融结，克保坚贞。且无瓦釜之鸣，是称土铏之德。器惭瑚琏，贡异砮丹。既尚质而为先，亦当无而有用。谨遣某官某乙随状封进。谨奏。"

In the eighth year of the Yuanhe reign of Emperor Xianzong (personal name Li Chun) of the Tang Dynasty (813), Yuancui, governor of Raozhou, who supervised the manufacture of the porcelain as tribute to the imperial court, asked Liu Zongyuan (an official and a scholar in the Tang Dynasty) to write a memorial to the Emperor with regards to the tributary porcelain wares.

Subject: Concerning this batch of tributary porcelain wares

Your majesty,

This batch of tributary porcelain wares from Raozhou is made with well selected raw materials of superior quality. In terms of shape, it can be regarded as a model of masterpieces. Not only does it inherit the superb craftsmanship of ancient potters, without any flaws, but also its firing process is almost perfect ensuring its firmness and fineness. In particular, the sound when touching it is totally different from that ordinary food utensils make, and its texture can be equivalent to the earthen bowl used by Yao and Shun. In terms of value, they are superior to the sacrificial utensils like Hulian which prove inferior by contrast, and they are almost as precious as with Huodan. With the combination of the virtuality and reality, it is far superior to others for its pursuit of elegance and nobility in simplicity. I would like to send an official named Yi to escort this batch of tribute porcelain wares enclosed with this memorial that will be presented to the throne.

Yours faithfully,

唐宪宗李纯之元和十一年（816 年），江州司马白居易在送友人到浔阳滋浦时，遇见商妇夜弹琵琶，遂作名篇《琵琶行》。诗中吟道："商人重利轻别离，前月浮梁买茶去。"此前，唐德宗李适之贞元十五年（799 年），白居易曾到浮梁县看望兄长白幼文，并寓居浮梁。唐宪宗李纯之元和十二年（817 年），白幼文卒于下邽，白居易写下《祭浮梁大兄文》，以表哀思。

In the eleventh year of the Yuanhe reign of Emperor Xianzong (personal name Li Chun) of the Tang Dynasty (816), when seeing a friend off at Pengpu, Xunyang, Bai Juyi, who was demoted to the assistant prefectship of Jiu Jiang, met a merchant woman playing the pipa at night. Afterwards he wrote in a renowned work "The Song of the Pipa Player", "the merchant cared for money much more than for me, one month ago he traveled to Fuliang to purchase tea." Earlier, in the

浮梁茶园
A Tea Plantation in Fuliang County

fifteenth year of the Zhengyuan reign of Emperor Dezong (personal name Li Shi) of the Tang Dynasty (799), Bai Juyi once paid a visit to his elder brother Bai Youwen in Fuliang County. Later he decided to live in Fuliang. In the twelfth year of the Yuanhe reign of Emperor Xianzong (personal name Li Chun) of the Tang Dynasty (817), Bai Youwen died at Xiagui town. Bai Juyi wrote a poem on "Sacrificing My Eldest Brother in Fuliang" to express his sorrow.

五代，景德镇瓷业状况尚未见古籍记载。

The manufacture of porcelain in Jingdezhen during the Five Dynasties has not been recorded in ancient books.

浮梁山涧

Mountain Stream in Fuliang County

但古窑址调查表明，五代时景德镇瓷业已初具规模，除生产青瓷外，还生产白瓷，质量为同期各窑口最优，白度值达到 70 以上，与现代瓷器几无差别。白瓷的面世，使景德镇成为中国南方最早生产白瓷的地区，结束了青瓷在南方一枝独秀的垄断地位，打破了唐代以来"南青北白"的生产格局。

However, the survey of ancient kiln sites shows that during the Five Dynasties, Jingdezhen porcelain industry have developed to a certain scale. Besides celadon, white porcelain was also produced in Jingdezhen. Its quality was the best among other kilns in the same period, and it was no different from modern porcelain, for its whiteness could reach more than 70%. With the invention of white porcelain, Jingdezhen became the first place to produce white porcelain wares in southern China, not only ending the monopoly of celadon in the south but also breaking the porcelain industry pattern of "green glazed wares of the South and white wares of Northern China" since the Tang Dynasty.

白瓷的创烧和发展意义重大。

The creation and development of white porcelain are of great significance.

中国瓷器可分为颜色釉和彩绘瓷两大类。白瓷的出现，不仅增加了一种美丽的釉色，还为彩绘瓷提供了理想的底色。元代以后，景德镇彩瓷逐渐成为主角。如果追根溯源，是五代白瓷为元明清彩瓷的发展铺平了道路。五代景德镇窑场不仅创烧出了白瓷，还初步创烧出早期青白瓷，成为宋代青白瓷的先导。

Chinese porcelain can be divided into two major categories, namely, color glazed porcelain and painted porcelain. The emergence of white porcelain not only adds a kind of beautiful glaze but also provides an ideal background for painted porcelain. Since the Yuan Dynasty, Jingdezhen colored porcelain has gradually become the main stream. To trace to its source, it was the white porcelain of the

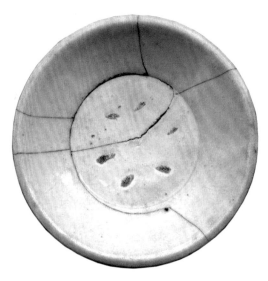

五代白瓷碟
A White-glazed Dish from the Five Dynasties

Five Dynasties that paved the way for the development of colored porcelain in the Yuan, Ming and Qing Dynasties. The period of the Five Dynasties witnessed the creation of white glazed porcelain wares as well as early greenish-white porcelain wares in Jingdezhen, which thereafter became the leader of greenish-white porcelain in the Song Dynasty.

第二编

成于宋元

II Maturing in the Song and Yuan Dynasties

景德置镇

The Establishment of Jingde Town

北宋太祖赵匡胤之开宝八年（975年），浮梁县属饶州，为望县，隶江东路。北宋真宗赵恒之景德元年（1004年），置景德镇，并设立监镇厅，仍辖于浮梁县。北宋神宗赵顼之元丰四年（1081年），浮梁县隶江南东路。

In the eighth year of the Kaibao reign of Taizu (personal name Zhao Kuangyin) of Northern Song Dynasty (975), Fuliang County belonged to Raozhou, which was Wangxian County, subordinate to Jiangdong Lu. In the first year of the Jingde reign of Emperor Zhenzong (personal name Zhao Heng) of the Northern Song Dynasty (1004), Jingdezhen was set up while a town supervision hall was established, which still was under the jurisdiction of Fuliang County. In the fourth year of the Yuanfeng reign of Emperor Shenzong (personal name Zhao Xu) of Northern Song Dynasty (1081), Fuliang County was subordinate to Jiangnan East Lu.

北宋景德元年（1004年），景德镇因皇帝年号得名。

Jingdezhen earned its name in the first year of the Jingde reign of the Northern Song Dynasty (1004).

浮梁红塔

The Red Tower in Fuliang County

《江西省大志·陶书》记载："陶厂。景德镇在今浮梁县西兴乡，水土宜陶。宋景德中，始置镇，因名。"《宋会要辑稿》则记载："江东东路饶州浮梁县景德镇，景德元年置。"以皇帝年号作为地名，宋代共有17例。史料还记载，山东郓州平阴县景德镇，北宋仁宗赵祯之景祐二年（1035年）置。至明代宗朱祁钰之景泰、孝宗朱祐樘之弘治年间（1450～1505年），改称安平镇。

As recorded in Taoshu, *Jiangxi Province General Gazetteer*, "Jingde town, known as 'Taochang' (ceramic factory), was located at Xixing Township, Fuliang County. The water and clay there are suitable for ceramics. Jingde town was first set up receiving its present name in the Jingde era during the Song Dynasty." According to *the Compilation of Song's Regulations*, "Jingde town, Fuliang County, Raozhou, Jiangdong East Lu, was set up in the first year of the Jingde reign." There were altogether 17 places given the title of the emperor's year in the Song Dynasty. In addition, according to historical records, Jingdezhen, Pingyin County, Yunzhou, Shandong Province, was established in the second year of the Jingyou reign of Emperor Renzong (personal name Zhao Zhen) of the Northern Song Dynasty (1035), and it was renamed as Anping town during the period between the Jingtai reign of Emperor Daizong (personal name Zhu Qiyu) and the Hongzhi reign of Emperor Xiaozong (personal name Zhu Youtang) of the Ming Dynasty (1450-1505).

《浮梁县志》记载："宋景德中，始置镇，因名。置监镇一员，以奉御董造。""宋真宗遣官制瓷，贡于京。"即遵照皇帝旨意，监烧宫廷用瓷。终宋一朝，景德镇青白瓷一直是宫廷用瓷的组成部分。随着瓷业生产的快速发展，一位监镇官已难以应付，后来增设窑丞之职以辅佐监镇官，专司窑业管理、课税、派烧和护送御器等职守。

According to *Fuliang County Chronicle*, "Jingde town was first set up in the

Jingde era during the Song dynasty. Afterwards, an official was assigned to the town to supervise the manufacturing of porcelain for the imperial court." "A court official was dispatched by the Emperor Zhenzong of the Song Dynasty to supervise the production of porcelain as tribute for the Capital." It means that the official supervised the production of court porcelain in accordance with the emperor's decree. In the late Song Dynasty, Jingdezhen greenish-white porcelain was always an integral part of court porcelain. With the rapid development of porcelain production, it was difficult for one supervisor to cope with the porcelain affairs. Later, the post of kiln official was added to assist the town supervisor, specializing in kiln management, levying of taxes, arranging firing and escorting imperial wares.

北宋仁宗赵祯之庆历五年（1045 年），景德镇窑丞齐护在护送御器过程中，因行从误毁御器而吞器身死。仁宗感其贤，于皇祐元年（1049 年）诏封他为新安元帅掣魔侯，立庙祭祀。齐护任窑丞 9 年，是史料可考的首位景德镇窑务官员。

In the fifth year of the Qingli reign of Emperor Renzong (personal name Zhao Zhen) of the Northern Song Dynasty (1045), Qi Hu, a kiln official of Jingdezhen, died of swallowing imperial porcelain due to an accidental destruction of the imperial wares caused by escorts. In 1049, Emperor Renzong, moved by his virtue, made an imperial edict to award him as "Xin'an marshal"-the demon hunter, and set up a temple to offer sacrifices to him. Qi Hu had worked as the kiln official for nine years, who was the first kiln official in Jingdezhen by verified historical data.

北宋神宗赵顼之元丰五年（1082 年），推行市易法的王安石纳余尧臣之策，设景德镇瓷窑博易务，专司瓷器贸易与税收，即以较便宜价格收购，以较高价格卖给商人，所收差价及商税作为国家财源上缴国库。宋代名窑众多，只有景德镇设此机构。

In the fifth year of the Yuanfeng reign of the Northern Song Dynasty (1082), Wang Anshi, who promoted "Market Exchange Law", adopted the measure proposed by Yu Yaochen, and set up Jingdezhen porcelain kiln affairs branch, specializing in porcelain trade and taxation, i.e., it was purchased at a cheaper price and sold to merchants at a higher price, and the price difference and the porcelain trade tax were turned over to the state treasury as a source of national wealth. There were many famous kilns in the Song Dynasty, but only Jingdezhen set up this institution.

进坑瓷石古道遗迹
The Relics of Ancient Porcelain Stone Road at Jinkeng

五代到北宋时期，景德镇窑场数目很多。《陶记》记载："景德陶，昔三百余座。""进坑石泥，制之精巧。湖坑、岭背、界田之所产，已为次矣。"考古发现，这一阶段 20 处古窑址均分布于南河流域，有 16 处终烧于北宋，延烧时间较长的是南市街和湖田窑，分别终烧于元代和明代中期。

During the period from the Five Dynasties to the Northern Song Dynasty, there were many kilns in Jingdezhen. It is recoded in the *Notes of Ceramics* that "All together there were over three hundred kilns in Jingdezhen." "The porcelain stone and clay made from Jinkeng village were with high quality, while those from Hukeng, Lingbei and Jietian were of inferior quality." According to the archaeological findings, twenty ancient kilns at this stage were distributed at the South River basin, and sixteen of them terminated firing during the Northern Song Dynasty. Nanshijie and Hutian Kilns had produced ceramics for a long time, which finally ceased firing during the Yuan Dynasty and the middle of the Ming Dynasty respectively.

从已发现宋代窑业堆积中，可以联想到当时景德镇"村村窑火、户户陶埏"的盛况。此时，景德镇瓷业生产仍属于私人作坊小规模生产，造坯和烧窑还未完全分开经营，但已开始脱离农耕家庭副业性质，出现新兴行业和技术分工。

From the uncovered potsherd piles and other kiln industry relics of the Song Dynasty, it can be assumed that the grand occasion of the time that "porcelain was fired in every household of every village" in Jingdezhen. At that time, Jingdezhen still adopted a small-scale private workshop pattern in the manufacture of porcelain. Clay body shaping and kiln firing had not been completely separated, but they had begun to break away from the sideline nature of farming families, emerging new businesses and technical division of labor.

景德镇的镇区也因瓷业发展而日渐兴旺。

千年古樟
A Thousand-year Old Camphor Tree

The urban area of Jingdezhen was also increasingly thriving due to the development of porcelain industry.

宋代的景德镇，沿着北至观音阁、南到小港嘴的昌江东岸，展开聚落式布局。聚落以渡口码头为起点，向周边及纵深平地和窑址延伸而形成。南宋时期，景德镇人口当在 5000 户以上。工匠增多，商贾往来频繁，便因此定居下来。

In the Song Dynasty, Jingdezhen developed a settlement layout along the east bank of the Chang River from the Guanyin Pavilion in the north to Xiaogangzui in the south, taking the ferry wharf as the starting point that extended to the periphery, the deep flat land and kiln sites. The population of Jingdezhen during the Southern Song Dynasty was estimated to be more than 5,000. As the number of artisans, craftsmen and merchants had been gradually increasing, many of them settled down in the town.

《景德镇市地名志》记载，在景德镇老街区当中，宋及宋以前有人集居的地名有 56 个，其中称为街的有 4 个，称为弄的有 22 个，称为其他名称的有 30 个。如半边街上有蓝家祠，原为宋代武将蓝魁所建宗祠，后便以此为地名。还有小港嘴，宋代有刘氏居住。十八桥，宋代有小溪，上有石板桥，古名登瀛桥，桥上阁中有十八学士像，宋以来一直是热闹集市，有"十八桥赛洛阳"之说。

According to the records in *Jingdezhen Gazetteer*, among the ancient blocks of Jingdezhen, there were fifty-six places' names inhabited in the Song Dynasty and before the Song Dynasty, including four places called streets, twenty-two places called Nong (lanes), and thirty others. For example, there was a Lan family temple on Banbian Street, which was originally built by Lan Kui, a military general of the Song Dynasty. Later, it was used as a place name. In addition, Xiaogangzui was a place where a Liu family lived in the Song Dynasty. There was a stream at Shibaqiao (eighteen bridges) in the Song Dynasty; over it was a slate bridge called Dengying bridge in ancient times. There were eighteen scholar statues in the pavilion on the bridge. There had been always a lot

春临水榭

A Waterside Pavilion in Spring

of hustle and bustle at the market since the Song Dynasty, which was said that "Shibaqiao could compete with Luoyang."

从此，景德镇逐步发展成为与湖北汉口镇、河南朱仙镇、广东佛山镇齐名的全国四大名镇之一，享有江南雄镇之美誉，镇上曾立有江南雄镇坊。

Since then, Jingdezhen has gradually developed into one of the four well-known towns in China, which is as famous as Hankou town in Hubei Province, Zhuxian town in Henan Province and Foshan town in Guangdong Province. Therefore, it enjoys the reputation of "the Great Town of Jiangnan." A memorial archway engraved with the Great town of Jiangnan was once erected in the town.

青白瓷系

Two Kiln Systems of Celadon and Greenish-White Porcelain

景德镇青白塔

The Qingbai Tower in Jingdezhen

宋代，被称作景德镇瓷器成功的时代。

The Song Dynasty is often called the successful era of Jingdezhen porcelain.

宋代经济发达，商业繁荣。宋人推崇理学，追求清白恬静的生活格调，对温润雅致的玉器充满向往，好玉成为美学趣味和社会风尚。五代晚期至北宋早期，在唐五代烧制青瓷和白瓷的基础上，景德镇模仿青白玉的色调与质感，创烧了外观和色泽类冰似玉的青白瓷。

The Song Dynasty was a splendid era with overwhelmingly developed economy and commerce. Neo-Confucianism appealed to people much more in the Song Dynasty, who tended to pursue an innocent and quiet lifestyle and yearn for gentle and elegant jade. Adoring jade became an aesthetic interest and social fashion. The period from the late Five Dynasties to the early Northern Song Dynasty witnessed the creation of greenish-white porcelain with a transparent jade-like green color in Jingdezhen by imitating the tone and texture of green and white jade based on the techniques of firing celadon and white porcelain of the Tang and Five Dynasties.

北宋青白釉台盏
A Qingbai Cup and Stand from Northern Song Dynasty

彭汝砺《送许屯田诗》说："浮梁巧烧瓷，颜色比琼玖。"

Peng Ruli, a writer of the Song Dynasty, once wrote in a poem titled "Say Farewell to Xu Tuntian," "Porcelain made at Fuliang was highly prized, and its color is better than Qiongjiu (Qiongjiu is the name of jade)."

《陶记》记载，景德陶"埏埴之器，洁白不疵。故鬻于他所，皆有'饶玉'之称"。青白瓷吸收南北窑口的形制款式及刻花装饰上的各种优点，造型上集纳金银器工整精致的特点，釉色青白淡雅，胎质坚致白细，常在器壁上雕有精美花饰，纹隙之处积釉呈色较深，映现出青色的纹影，有"青如天、白如玉、明如镜、薄如纸、声如磬"之誉。入清以后，青白瓷改称影青瓷。

According to the *Notes of Ceramics*, "Porcelain wares made in Jingdezhen were white, clean and flawless. Wherever they were sold in other provinces in China, people called them Raoyu (Jade of Rao)." Greenish-white porcelain inherits the shape and style of the north and south kilns and collects widely all good qualities in carved decoration. In terms of modeling, it integrates the characteristics of neat and exquisite gold and silver wares. Greenish-white porcelain pieces are characterized by elegantly bluish glaze, fine-textured and white body, exquisite flowers patterns carved on the surface where the accumulated glaze embedded in the striated gaps appear dark, casting a bluish shadow. Therefore, greenish-white porcelain has long been known "as blue as the sky, as white as jade, as bright as a mirror, as thin as paper and as sound as a bell." After entering the Qing Dynasty, greenish-white porcelain was renamed as shadow celadon.

宋景德镇青白釉孩儿枕

A Head Rest with Qingbai Glaze, Jingdezhen, Song Dynasty

李清照《醉花荫》说："佳节又重阳，玉枕纱厨，半夜凉初透。"

Li Qingzhao once wrote in a poem titled "Drunk in the Shade of Flowers" that "A holiday once again, it is the Double Ninth; through embroidered pillows and gauze veils, in the middle of the night, coolness begins to come through."

青白瓷深受人们的喜爱和社会欢迎，成为十分畅销的商品。《景德镇陶录》称："其器尤光致茂美，当时则效，著行海内，于是天下咸称景德镇瓷器。"宋元时期，景德镇青白瓷产量为全国之冠，流通地域达到全国三分之二的省份。除个别为本省产品外，绝大多数为景德镇窑生产，以至于全国"舟辇所达、无非饶器"。

Greenish-white porcelain has become a very popular commodity, for it is deeply loved by people and welcomed by the society. According to *Records of Jingdezhen Ceramics*, "Jingdezhen porcelain was featured with good luster, high density and rich decorative effects. At that time, other places began to imitate its techniques for making porcelain. As porcelain was exported at home and abroad, it was recognized by the world." During the Song and Yuan Dynasties, Jingdezhen greenish-white porcelain ranked first in overseas porcelain output, and its circulation area reached 2/3 provinces in China. Except for a few local products, the vast majority of porcelain wares were made from Jingdezhen kilns, which were transported to all over the country by land and water.

唐代中后期至宋代开启的海上丝绸之路，又称陶瓷之路。

The Maritime Silk Road, known as "the Porcelain Road", was developed from the middle and late Tang Dynasty to the Song Dynasty.

陶瓷之路上，撒落了无数的中国各朝瓷器，数量最大的是景德镇瓷器。宋元外销瓷主要有景德镇窑系的青白瓷、龙泉窑系的青瓷、建窑系的黑釉

南宋青白釉印花双凤纹碗

A Qingbai Bowl Incised with Two Phoenixes, Southern Song Dynasty

瓷，数量最多的是青白瓷。由于青白瓷供不应求，在通往沿海港口的线路上，出现了许多纯以外销为目的的仿烧民窑，以至于在东南沿海地区形成庞大的青白瓷窑系。

Among countless Chinese porcelains of various dynasties scattered along the ancient "Porcelain Road", the vast majority of these objects were made in Jingdezhen. The export porcelain of the Song and Yuan Dynasties mainly included greenish-white porcelain from Jingdezhen kiln system, celadon porcelain from Longquan kiln system and black glazed porcelain of Jianyao kiln system, among which greenish-white porcelain took the largest proportion. Because the supply of greenish-white porcelain was in short supply, many imitating private kilns emerged along the coastal ports for the sake of export porcelain, resulting in the formation of a huge greenish-white porcelain kiln system in the southeast coastal area.

在宋代八大瓷系中，青白瓷窑系居于首位。

Among the eight great kiln systems during the Song Dynasty, bluish-white porcelain kiln system ranked first.

湖田窑遗址
The Hutian Ancient Kiln Site

景德镇青白瓷经历了北宋太祖赵匡胤之建隆元年（960 年）至真宗赵恒之乾兴元年（1022 年）的创烧期、仁宗赵祯之天圣元年（1023 年）至钦宗赵桓之靖康二年（1127 年）的繁荣期、南宋高宗赵构之建炎元年（1127 年）至恭帝赵㬎之德祐二年（1276 年）的发展繁荣期、恭帝赵㬎之德祐二年（1276 年）至元末（1368 年）的衰落期四个发展阶段。

Jingdezhen greenish-white porcelain experienced four development stages, i.e., the first stage was the initial period from the first year of the Jianlong reign (960) of Emperor Taizu (personal name Zhao Kuangyin) of the Northern Song Dynasty to the first year of the Qianxing reign of Emperor Zhenzong (personal name Zhao Heng) of the Northern Song Dynasty (1022); the second stage was the prosperous period from the first year of the Tiansheng reign (1023) of Emperor Renzong (personal name Zhao Zhen) to the second year of the Jingkang reign (1127) of Emperor Qinzong (personal name Zhao Huan); the third stage was the most prosperous period from the first year of the Jianyan reign (1127) of Emperor Gaozong (personal name Zhao Gou)to the second year of the Deyou reign (1276) of Emperor Gongdi (personal name Zhao Xian) of the Southern Song Dynasty;

the last stage was the decline period from the second year of the Deyou reign (1276) of Emperor Gongdi (personal name Zhao Xian) to the end of the Yuan Dynasty (1368) .

元代中晚期，青白瓷走向衰落，逐渐被青花瓷、釉里红和卵白釉瓷取代。

During the middle and late Yuan Dynasty, bluish-white porcelain was in decline, which was gradually replaced by blue and white porcelain, underglaze red and egg white glaze porcelain.

浮梁瓷局

Fuliang Porcelain Bureau

　　元代，浮梁县仍属饶州路，归于江浙行中书省管辖。元世祖孛儿只斤·忽必烈之至元十五年（1278 年），在景德镇设立浮梁瓷局。成宗孛儿只斤·铁穆耳之元贞元年（1295 年），浮梁升县为州，隶属于饶州路。

浮梁古村落

An Ancient Village in Fuliang

During the Yuan Dynasty, Fuliang County still belonged to Raozhou Lu, which was under the jurisdiction of Jiangzhe Xingsheng (Zhejiang branch secretariat). In the fifteenth year of the Zhiyuan reign of the Yuan Dynasty (1278), Emperor Kublai Khan Borjigin established Fuliang Porcelain Bureau in Jingdezhen. In the first year of the Yuanzhen reign (1295) of Emperor Cheng Zong (personal name Timur Borjigi), Fuliang County was upgraded to Zhou, belonging to Raozhou Lu.

《元史》记载："浮梁瓷局，秩正九品，至元十五年立。掌烧造瓷器，并漆造马尾棕藤笠帽等事。大使、副使各一员。"浮梁瓷局隶属将作院，虽然除掌烧造宫廷所需的瓷器外，还要掌管马尾棕、藤笠帽等生产事宜，但其主责是瓷。这是自古代至元代时，中国唯一的瓷局。

According to *the History of the Yuan*, "Fuliang Porcelain Bureau, ranking 9A, was established in the fifteenth year of the Zhiyuan reign of the Yuan Dynasty, which was responsible not just for porcelain manufacturing but for the production of coir ropes, rattan hats and other products. Two officials were dispatched separately as an envoy and a deputy envoy." Fuliang Porcelain Bureau was subordinate to the Imperial Manufactories Commission, whose functions were to serve as supervision and management of porcelain production in Jingdezhen for official use and other products, the main responsibility of which was porcelain production. This was the only porcelain bureau in China from ancient times to the Yuan Dynasty.

浮梁瓷局并非专门生产御器的场地，而是官府对窑场进行征派的机构。其职能主要是负责向各窑场推派、收购、输送、检查等，即从民窑产品中百中选一、千中选十，挑选烧造精良的瓷器作为贡品，"有命则贡，无命则止"。

Fuliang Porcelain Bureau was not only a place specializing in the manufacture of imperial wares, but an organization for the government to levy and dispatch kilns. Its function was mainly responsible for apportion, purchasing, transporting and inspecting the kilns, i.e., strictly selecting the most perfect porcelain wares from numerous private kiln products as tribute. These imperial kilns were normally operated only with the approval of the Son of the Heaven, otherwise it would cease operation.

元青白釉双耳瓶
A Qingbai Vase with Double Ears, Yuan Dynasty

元代，景德镇制瓷业进入黄金时期。

In the Yuan Dynasty, Jingdezhen porcelain industry entered the golden age.

在中国瓷器发展史上，元代属于承前启后的重要时期。宋金时代的南北各地主要瓷窑，如钧窑、磁州窑、定窑、吉州窑、德化窑，以及山西地区的黑瓷和南方各地的青白瓷，在元代虽然继续生产，但到了元代后期，很多瓷窑都已成强弩之末，能代表瓷器生产时代特点的只有景德镇窑。

The Yuan Dynasty is a transitional age in the development of Chinese porcelain. In the Song and Jin Dynasties, the main kilns in the north and south, such as Jun kiln, Cizhou kiln, Ding kiln, Jizhou kiln and Dehua kiln, as well as kilns for black porcelain in Shanxi and greenish-white porcelain in the south, continued operating in the Yuan Dynasty, but many of them finally declined in the late Yuan Dynasty. Only Jingdezhen kilns represented the characteristics of the porcelain production era.

元初，景德镇青白瓷生产沿袭南宋时期瓷器的造型、装饰、胎釉特征，但不久即发生变化。从传世及出土的瓷器看，元代景德镇除继续烧制青白瓷、白瓷、黑瓷外，还创烧了具有划时代意义的新品种青花瓷、釉里红瓷、卵白釉枢府瓷，以及红釉、蓝釉等高温颜色釉瓷，孔雀绿等低温颜色釉瓷，开创了彩瓷新时代，推动景德镇逐渐成为中国瓷器的生产中心。

In the early Yuan Dynasty, the manufacture of greenish-white porcelain carried on the legacies of the Southern Song Dynasty. The characteristics of modeling, decoration and body glaze were the same as those of the previous dynasty, but things changed thereafter. Judging from the porcelain objects handed down and unearthed, Jingdezhen in the Yuan Dynasty, besides continuing to make greenish-white porcelain white porcelain and black porcelain, created new varieties with epoch-making significance, including blue and white porcelain, underglaze red porcelain, Shufu glaze porcelain, high-temperature glazed porcelain represented by red glaze and blue glaze), and low-temperature glazed porcelain represented by peacock green, launching a "new era" for colored porcelain, which promoted Jingdezhen to gradually become the manufacturing center of Chinese porcelain.

湖田窑遗址展厅
The Exhibition Hall of the Hutian Ancient Kiln Site

元代疆域辽阔，欧亚之间诸多区域文明相互影响。蒙古族文化、伊斯兰文化、汉族传统文化、藏传佛教文化、欧洲基督教文化、高丽文化等多元并存，文化艺术呈现五彩斑斓的面貌。元统治者尊藏传佛教为国教，在饮食起居、服饰制度、工艺爱好上大多保留了民族特色，这种特色文化广泛渗透到工艺美术中，突出表现为尚白和尚蓝。由此，青花瓷和卵白釉瓷受到青睐。

As the Yuan Dynasty had a vast territory, many regional civilizations between Europe and Asia influenced each other. The co-existence of various cultures, such as Mongolian culture, Islamic culture, Han traditional culture, Tibetan Buddhist culture, European Christian culture and Korean culture, showed colorful allure in the field of culture and art. The rulers of the Yuan Dynasty respected Tibetan Buddhism as the national religion, and most of them retained their national characteristics in food and daily life, clothing system and craft hobbies. This characteristic culture widely penetrated into the arts and crafts, which were mainly reflected in white worship and blue admiration. Therefore, blue and white porcelain and egg white glaze porcelain became increasingly popular.

典型的元代瓷器具有胎厚、硕大、体重的时代特点。

Typical porcelain wares of the Yuan Dynasty are marked by thick body, large in size and heavy weight.

在工艺上，从南宋至元代，高岭土首次在景德镇用作瓷胎原料，独创单一瓷石到瓷石加高岭土的二元配方，大大降低了变形率，成品率大幅上升，瓷质白度提高，开启了高温硬质瓷的新纪元。

In terms of the manufacturing process, kaolin was first used as raw material for porcelain body in Jingdezhen during the period from the Southern Song Dynasty to the Yuan Dynasty. "Two element based formula" consisting porcelain stones and kaolin was created instead of single porcelain stone recipe, which greatly reduced the deformation rate, increased the yield, and improved the whiteness of porcelain, opening a new era of high-temperature hard porcelain.

元代海外贸易较宋代有所扩大，元代瓷器在东南亚地区出土的数量也大大超过了宋代。元代输出的瓷器主要是东南沿海地区瓷窑烧制的，除浙江龙泉窑青瓷、江西景德镇青白瓷外，浙江、福建地区大量瓷窑烧造的仿龙泉瓷与青白瓷也占有很大比重。元代后期，景德镇青花瓷也输往海外。

Compared with that of the Song Dynasty, the overseas trade of the Yuan Dynasty slightly expanded, and the number of porcelains unearthed in Southeast Asia of the Yuan Dynasty also greatly exceeded that of the Song Dynasty. The exported porcelain in the Yuan Dynasty was mainly fired by porcelain kilns in the

作为外销瓷的元青花大盘
A Large Export Porcelain Blue and White Plate, Yuan Dynasty

southeast coastal areas. In addition to the celadon of Longquan kiln in Zhejiang and greenish-white porcelain of Jingdezhen in Jiangxi, a large number of copies of Longquan porcelain and greenish-white porcelain from Zhejiang and Fujian also accounted for a large proportion. During the late Yuan Dynasty, Jingdezhen blue and white porcelain was also exported overseas.

青花瓷器

Blue and White Porcelain

青花瓷是中国瓷器的主流品种之一。

Blue and white porcelain is one of the mainstream varieties of Chinese porcelain.

青花瓷韵

Passion for Blue and White Porcelain

青花瓷属于釉下彩瓷，是以含氧化钴的钴矿为原料，在瓷坯上描绘纹饰，再罩上一层透明釉，经高温还原焰一次烧成。钴料烧成后呈蓝色，具有着色力强、发色鲜艳、烧成率高、呈色稳定的特点。

Blue and white porcelain belongs to underglaze color porcelain. Blue and white porcelain is contrived by using cobalt ore containing cobalt oxide as raw material, which is painted on ware body, and then covered with a layer of transparent glaze, and fired in a high temperature reduction flame. After firing, cobalt material turns blue. Blue and white porcelain is marked by strong coloring strength, excellent brightness, high firing rate and stable color.

成熟的青花瓷出现于元代景德镇湖田窑。

Mature blue and white porcelain appeared in the Hutian kiln in Jingdezhen in the Yuan Dynasty.

元青花菊纹诗文高足杯
A Blue and White Stem Cup with
Chrysanthemum Design and a Poem,
Yuan Dynasty

青花瓷的出现，结束了中国瓷器以单色釉为主的局面，把瓷器装饰推进到釉下彩的新时代，形成鲜明的中国瓷器特色。经由 14 世纪二三十年代创烧，到 15 世纪前期的明代，仅经过 70 年，景德镇青花瓷就占据了中国瓷器生产的主流。自此，景德镇青花瓷盛开 800 年不败，佳品迭现。其他大部分古老的瓷器窑场相形见绌，景德镇也由此成为中国的瓷都。

The emergence of blue and white porcelain ended the situation that Chinese porcelain was dominated by monochrome glaze, promoting porcelain decoration to a new era of underglaze color decoration and forming distinct Chinese porcelain characteristics. It took only 70 years of development-from the initial period of development （1320s-1330s） to the early Ming Dynasty in the early fifteenth century-that Jingdezhen blue and white porcelain had become the mainstream of Chinese porcelain production. Since then, Jingdezhen blue and white porcelain had enjoyed the reputation of "ever-green blue and white porcelain" for 800 years, continuously producing numerous masterpieces. As other ancient porcelain kilns gradually declined, Jingdezhen therefore became China's porcelain capital.

青花瓷创造于元代的景德镇，这是历史机缘。

Blue and white porcelain was created in Jingdezhen in the Yuan Dynasty, which was a historical opportunity.

元青花云龙纹带盖梅瓶

A Blue and White Meiping Vase and Cover with Dragon among Clouds, Yuan Dynasty

蒙古人站在草原上，抬头仰望，白云蓝天。到了冬季，俯视四野，皑皑白雪，一望无际，所以他们无比敬畏白与蓝两种色彩。尚白与崇蓝的文化对瓷器的影响，催生了以白色为底色、以蓝色为主色的青花瓷。海陆四通八达，钴原料苏麻离青从西亚进入中国，烧造的青花瓷又运回西亚各地。景德镇的能工巧匠抓住了历史的机缘，促成了青花瓷的流行。

The Mongols live on the grassland and look up at bright white clouds on a clear blue sky. When it is in winter season, there is snow everywhere stretching as far as the eye can see, so they are in incomparably awe of the two kinds of colors, namely, white and blue. The influence of the culture of white worship and blue admiration on porcelain gave birth to blue and white porcelain decorated with white as the background color and blue as the main color. As the sea and land extended in all directions, Sumaliqing, a special kind of imported blue and white cobalt that was used as decoration, entered China from West Asia; meanwhile, blue and white porcelain wares after being fired were transported back to all parts of West Asia. Skilled craftsmen in Jingdezhen triggered the popularity of blue and white porcelain by seizing the opportunity of history.

景德镇青花瓷的大量生产，主要是满足外贸市场的大量需求。当时，元代宫廷用瓷和国内市场需求，远不及官营和民营海外贸易的出口量，而且销往西亚、东非等地的产品质量最好，以至于超过了宫廷用瓷。

The massive production of Jingdezhen blue and white porcelain was mainly to meet the large demand of the foreign trade market. At that time, the demand for porcelain for the imperial court and the domestic market in the Yuan Dynasty were far less than the export volume of official and private overseas trade. Moreover, the porcelain sold to West Asia, East Africa and other places was of the best quality, even exceeding that of imperial porcelain.

海外贸易的高额利润促进了元代景德镇青花瓷的生产发展。元青花瓷窑址不仅在距镇区约 5 公里的湖田和浮梁县旧城出现，而且沿昌江东岸，北自观音寺，南到小港嘴，绵延几千米，在东西前街、后街之间宽 1 公里多的镇区内，窑址一个接一个。可以想象，当时的景德镇，来自波斯的商人与大元的官商、民商走街串巷，大量收购青花瓷。来自八方的官匠、民匠与作坊主日夜忙碌，加紧生产，窑场呈现"火光烛天，夜不能寝"的繁华景象。

元代馒头窑
The Mantou Kiln from the Yuan Dynasty

The high profits of overseas trade contributed to the production and development of Jingdezhen blue and white porcelain in the Yuan Dynasty. Yuan blue and white porcelain kiln sites appeared not only in Hutian and the old town of Fuliang County, which were about 5km away from the town, but also along the east bank of the Chang River, from Guanyin Temple in the north to Xiaogangzui in the south, stretching for several kilometers. Kilns were concentrated one after another about more than 1km wide between east and west front and back streets across the town. It can be imagined that in Jingdezhen at that time, merchants from Persia thronged the alleys and warehouses, mingling with a handful of officials and civilians from "Dayuan (the Great Yuan Dynasty)", purchasing blue and white porcelain in large amounts. Officials, craftsmen, folk craftsmen and workshop owners from all directions were working around the clock in the manufacture of porcelain, presenting a prosperous scene of "stayed up all night with the thundering of tens of thousands of pestles pounding the ground and the glare from the kiln fires lighting the sky" in the kiln factories.

此时，湖田等处仍以深层瓷石制瓷，品质低劣，高岭土又运入困难，力资昂贵，竞争力明显衰微终至没落。大批专业窑户争相向河洲集结。这不但促进了镇市的繁盛，更促成瓷业不再作为农耕文明的附庸，而向专业化、集约化、商品化的工业文明大步迈进。

In Hutian kiln and some other kilns, porcelain stone from deep below the ground was still used as the major material that contributed to poor quality porcelain products. Together with the difficulty of transporting kaolin into that area and the high cost of labor, Hutian kiln declined and eventually fell into ruins. By then, massive specialized kilns were scrambling to concentrate at the alluvial basin along the river, which not only fostered the prosperity of the town, but also turned the ancient town from being dominated by agricultural civilization to industrial civilization of specialization, intensification and commercialization.

元青花云龙牡丹纹大罐

A Large Blue and White "PEONY" Jar with Dragon among
Clouds, Yuan Dynasty

　　青花瓷因其所具有的历史和现实地位，已成为国际视野里的中国元素，中国文化里的显著标志。从历年国际上的天价拍卖行情中就可以窥见其价值：2005 年元青花鬼谷子下山图罐以 2.3 亿元人民币的天价拍出；2007 年元青花龙纹四系扁瓶以 9790 万元人民币成交；2012 年元青花鱼藻纹折沿盘以 6888.5 万元人民币成交；2011 年元青花龙纹大罐以 5376 万元人民币成交。

　　Because of its historical and current status, blue and white porcelain has become a Chinese element in the international perspective and a significant symbol in Chinese culture. Its value can be seen in a glance from the high price auction market in the world over the years. The typical cases are as follows: On July 12, 2005, a Yuan Dynasty blue and white vase with the scene of "Gui Guzi Descends the Mountain" was sold at the high price of 230 million CNY. In 2007, a blue and white porcelain flat vase with four rings was sold for 97.9 million CNY; In 2012, a Yuan Dynasty blue and white folded rim plate with fish algae pattern was sold for 68.885 million yuan; In 2011, a Yuan Dynasty large blue-and-white jar with dragon pattern was traded with 53.76 million CNY.

欧洲人曾认为，青花瓷是中国人送给欧洲文艺复兴的礼物。

Europeans once believed that blue and white porcelain was a gift from the Chinese to the European Renaissance.

第三编　盛于明清

III　Attaining Its Acme in the Ming and Qing Dynasties

多彩时代
The Colorful Era

　　明太祖朱元璋之洪武元年（1368 年），饶州路改鄱阳府，浮梁州属之，洪武二年（1369 年）改浮梁州为县，属饶州府。洪武十年（1377 年）置行省，浮梁隶江西布政使司，属九江道饶州府。清代初期裁并各道，浮梁县隶

龙珠阁
The Longzhu Pavilion

属江西省。

In the first year of the Hongwu reign of Emperor Taizu (personal name Zhu Yuanzhang) of the Ming Dynasty (1368), Raozhou Lu was changed to Poyang Prefecture with Fuliang under its jurisdiction. In the second year of the Hongwu reign (1369), Fuliang District became Fuliang County, belonging to Raozhou Prefecture. In the tenth year of the Hongwu reign (1377), as the Branch Secretariats (Xingsheng) were set up, Fuliang was subordinate to the Jiangxi Buzhengshisi (provincial administrative commission), belonging to Raozhou Prefecture, Jiujiang Dao. In the early Qing Dynasty, as administrative regions were redistributed and merged, Fuliang County was subordinate to Jiangxi Province.

沈嘉徵《窑民行》说："工匠来八方，器成天下走。"

According to Shen Jiazheng's poem titled "Yaominxing", "To it craftsmen, come from four directions; from it, vessels go to all parts of the world."

景德镇自古是一座移民之城。明代中期以后，景德镇瓷器畅销海内外，吸引了江西境内其他地区一些破产和贫困农民到此地谋生，为瓷器大量生产提供了源源不断的劳动力。外地各省的商贩也纷纷向景德镇会聚。到清末，有全国 18 省 68 县的 20 余万人在景德镇工作和生活，外来移民及其后代占总人口的 95%以上，浮梁本地人占比很小，是全国少有的移民大大超过县域人口的市镇。

Jingdezhen has been a city of immigrants since ancient times. After the mid Ming Dynasty, porcelain wares made in Jingdezhen were well sold at home and abroad, which attracted some bankrupt and poor farmers from other parts of Jiangxi to make a living here, providing a steady stream of labor force for porcelain mass production. In addition, vendors from other provinces also gathered in Jingdezhen. By the end of the Qing Dynasty, more than 200,000 people from 68 counties in 18 provinces worked and lived in Jingdezhen. Immigrants and their descendants accounted for more than 95% of the total population, and local Fu-liang people accounted for very little. Therefore, it is a rare city in China where immigrants greatly exceed the native population of the county.

景德镇徽商老宅
A Former Residence of a Huizhou Merchant in Jingdezhen

清代以后，景德镇形成三大商帮。

After the Qing Dynasty, Jingdezhen formed three business groups.

商业为徽商经营，还垄断金融业，称徽帮。圆器业和窑业由都昌人经营，称都帮。琢器业、红店，以及与瓷业有关的服务业和其他大小行业，由抚州帮、南昌帮、丰城帮、吉安帮、奉新帮、饶州帮、安仁帮和各地旅景瓷商 26 帮经营，称杂帮，其中以抚州籍琢器帮为大帮。清末民初，三帮联合成立景德镇总商会。

Commercial business was run by Huizhou merchants who monopolized the financial trade, known as Huizhou Bang. Round wares and kiln business were controlled by Duchang people, known as Du Bang. The carving industry, hollow wares and Hongdian (colored drawing stores), as well as the service industry related to the porcelain industry and other large and small industries were operated by the so-called 26 miscellaneous Gangs (mixed Gang) including Fuzhou Gang, Nanchang Gang, Fengcheng Gang, Ji'an Gang, Fengxin Gang, Raozhou Gang, Anren Gang and the resident porcelain merchants in Jingdezhen, of which Fuzhou hollow wares Gang is the largest. At the end of the Qing Dynasty and the beginning of the Republic of China, these three groups jointly established Jingdezhen General Chamber of Commerce.

郑凤仪《浮梁竹枝词》说："夜阑惊起还乡梦，窑火通明两岸红。"

Zheng Fengyi wrote in a poem titled "Fuliang Bamboo Branch Song", "At night, the flames from the burning kilns are lighting up the night sky on both sides of the Chang River, which often awakens travelers from their dream of returning home."

由宋代的百花争艳，经元代的过渡，到了明代，中国瓷器形成景德镇一花独放的局面，代表时代特征的是景德镇瓷器。明代中期以后，景德镇瓷器占据全国主要市场，宫廷所用瓷制品也主要由景德镇供应。

The manufacture of porcelain in the Song Dynasty revealed dazzling exuberance, and the Yuan Dynasty was a transitional age in the development of Chinese porcelain. Jingdezhen had a monopoly on porcelain production in China until the Ming Dynasty, which represented the characteristics of porcelain industry at the times. Since the middle Ming Dynasty, Jingdezhen porcelain had occupied the main market in China, and porcelain wares for the imperial court were mainly supplied by Jingdezhen.

明代葫芦窑
A Gourd Shaped Kiln of the Ming Dynasty

龚鉽《陶歌》说："白胎烧就彩红来，五色成窑画作开。"

Gong Shi wrote in a poem titled "Tao Ge", "In the high-temperature firing, the glaze in porcelain body blends and changes naturally to form a variety of graphics, just like the freehand brushwork in the art of traditional Chinese painting, with both form and spirit, like truth and illusion, and infinite artistic conception."

明代宫廷在景德镇设立御器厂，不惜代价，促使景德镇制瓷业不断扩大新品种，提高产品质量，带动了民窑的进一步发展。民窑在扩大市场的基础上，也精益求精。到嘉靖以后，宫廷所需御器中的钦限瓷器，已由民窑官古器户烧造。民间的中上阶层人家，普遍使用景德镇民窑瓷器，特别是青花瓷。

The court of the Ming Dynasty set up the imperial ware factory in Jingdezhen, which facilitated Jingdezhen porcelain industry to continuously create new varieties and improve product quality at all costs; meanwhile, it also drove the further development of folk kilns. On the basis of expanding the market, folk kilns also strived for perfection. After Jiajing's reign, private kilns that assisted with imperial production were permitted to produce the imperial wares that were used exclusively by the emperor and the court. Porcelain wares, especially blue and white porcelain, made in civilian kilns in Jingdezhen, were widely used by the upper middle-class families in civil society.

明清两代，彩瓷全面发展，令景德镇瓷业独步天下。

During the Ming and Qing Dynasties, the all-round development of colored porcelain made Jingdezhen porcelain industry unparalleled in the world.

入明以后，景德镇瓷业生产规模逐步超过前代，制瓷工艺有了一系列创新，从单色釉到多色釉，由釉下彩到釉上彩，青花之外又出现了红绿彩、

明宣德青花云龙纹蟋蟀罐

A Blue and White Cricket Jar with Dragon
among Clouds, Xuande Period, Ming
Dynasty

五彩、素三彩、青花斗彩等全新装饰技法，开启了中国瓷器的多彩时代。

Since entering the Ming Dynasty, the production scale of Jingdezhen porcelain industry had gradually exceeded that of the previous generation. Moreover,
there were a series of innovations in the manufacturing process of porcelain. Decoration techniques were continuously updated from monochrome glaze to mul-
ti-color glaze, from underglaze color to overglaze color. Besides blue and white
decoration, some completely new decoration techniques emerged, including reddish green, polychrome, plain tricolor, blue and white doucai and so on, which
opened the colorful era of Chinese porcelain.

英国学者李约瑟认为，景德镇是世界上最早的工业城市。

Joseph Needham, a British scholar, insists that "Jingdezhen was the earliest
industrial city in the world."

明代，景德镇瓷业生产进入了城镇化工场手工业阶段，其产业规模、
专业分工、运行管理、生产关系，已显露出近代工业的雏形，出现了资本

主义生产关系的萌芽，成为中国制瓷中心和世界制瓷工业大都会。延续至清代，景德镇制瓷进入了独领风骚600多年的兴盛时期。

In the Ming Dynasty, the manufacture of porcelain in Jingdezhen entered the handicraft stage of urban industry. In terms of its industrial scale, professional division of labor, operation management and production relations, Jingdezhen had already had the embryonic form of modern industry and the germination of capitalist production relations. Hence, Jingdezhen became the center of China's porcelain manufacturing and the metropolis of the world's porcelain industry. Until to the Qing Dynasty, Jingdezhen had entered into a period of prosperity when it took the lead in the manufacture of porcelain, lasting for more than 600 years.

明神宗朱翊钧之万历二十七年（1599年），江西矿税使兼理景德镇窑务太监潘相督造龙缸，因器大难成，累不完工，民受鞭，或苦饥羸，陶人

佑陶灵祠

The Temple of Blessing Ceramics

童宾至以身赴火，罹其凶毒。瓷工崇敬童宾，尊为风火仙，建庙奉祀，清廷敕封广利窑神。

In the 27th year of the Wanli reign of Emperor Shenzong (personal name Zhu Yijun) of the Ming Dynasty (1599), Pan Xiang, a court eunuch, working as the mineral tax envoy of Jiangxi Province and concurrently in charge of Jingdezhen kiln affairs, supervised to the construction of a dragon vessel. Because the size of vessel is prodigious, it could not be finished in time, so the potters were mercilessly abused and persecuted for this. Finally, in a gesture of protest, a pottery man named Tong Bin, leapt into the kiln, accelerating the combustion with his own flesh and bones. His fellow potters revered Tong Bin as the "Immortal of the Flaming Kiln", and built a temple to him. In addition, the Qing Government bestowed the title "Guangli Kiln God" to him.

清穆宗爱新觉罗·载淳之同治八年（1869 年），德国地质学家李希霍芬

高岭瓷土矿遗址

The Relic Site of Kaolin Clay Mines

到景德镇考察后，按高岭读音译成 Kaolin 一词，成为国际通用名词，浮梁县高岭村也由此成为高岭土命名地。

In the eighth year of the Tongzhi reign of Emperor Muzong (personal name Aixinjueluo Zaichun), of the Qing Dynasty (1869), the German geologist Ferdinand von Richthofen visited Jingdezhen to study porcelain and translated the Chinese "Gaoling" ("Gaoling" in Pinyin) as "Kaolin", which became an international common terminology. Therefore, Gaoling village in Fuliang County became the naming place of kaolin.

清代，景德镇始终保持中国瓷都的地位。

In the Qing Dynasty, Jingdezhen always maintained the status of China's porcelain capital.

清康熙五彩花鸟纹尊

A Wucai Zun with Birds and Flowers, Kangxi Period, Qing Dynasty

清代，除了宫廷用瓷，民间用瓷也绝大部分由景德镇供应。尤其是清代前期的康、雍、乾三朝，景德镇制瓷在工艺技术和产量上，都达到历史高峰，技艺更加娴熟精湛，品种更加丰富多彩。康熙时期（1662~1722 年）的青花瓷、五彩瓷、郎窑红、豇豆红、珐琅彩瓷，雍正时期（1723~1735 年）的粉彩瓷、高温窑变釉，乾隆时期（1736~1795 年）的镂雕瓷、仿生瓷等，均为集历代南北名窑之大成。

In the Qing Dynasty, besides porcelain for the imperial court, a majority of folk porcelain wares were also supplied by Jingdezhen. Especially during the reigns of Emperor Kangxi, Yongzheng, and Qianlong in the early Qing Dynasty, Jingdezhen porcelain experienced a historical peak in process techniques and output, with more exquisite skills and more colorful varieties. Blue and white porcelain, multicolored porcelain, polychrome porcelain, Lang kiln red glaze porcelain, cowpea red and enamel colored porcelain of the Kangxi reign（1662-1722）, famille rose porcelain and high temperature kiln transformation glaze porcelain of the Yongzheng reign（1723-1735）, carved porcelain and bionic porcelain of the Qianlong reign（1736-1795）and so on are all a collection of famous kilns in the north and south of the past dynasties.

乾隆时期（1736~1795 年）以后，伴随国运衰退，景德镇制瓷业逐渐衰落。

Since the Qianlong reign（1736-1795）, with the decline of national fortune, Jingdezhen porcelain industry had also gradually declined.

天工开物

Exploitation of the Works of Nature (Tiangong Kaiwu)

张宿煌《景德镇竹枝词》说："十万人烟背枕河，火龙盘踞起窑窝。"

Zhang Suhuang wrote in a poem titled "Jingdezhen Bamboo Branch Song", "Hundreds of thousands of people who live along the Taoyang Shisan Li (thirteen miles) road saw jets and flashes of red fire spray from the chimney of the kilns at nights."

制瓷古作坊

An Ancient Porcelain Manufacturing Workshop

自明神宗朱翊钧之万历年间（1573～1620年），湖田窑熄灭了近700年的窑火之后，延续到清代，景德镇瓷业生产已基本集中于市镇，标志着制瓷业作为农耕经济附庸时期的彻底终结，在全国率先跃入了专业化、集约化、规模化、商品化生产的手工业文明的新时代，也带动景德镇日益繁荣。

During the Wanli reign of Emperor Shenzong (personal name Zhu Yijun) of the Ming Dynasty (1573-1620), Hutian kiln extinguished the kiln fire after burning for nearly 700 years. Until the Qing Dynasty, Jingdezhen porcelain production was basically concentrated at the town, marking the complete end of the period when the porcelain industry was a vassal of the agricultural economy, and it began to take the lead in jumping into a new era of handicraft civilization of professional, intensive, large-scale and commercialized production in the country, which also drove the growing prosperity of Jingdezhen.

到清末，景德镇制瓷业已形成非常完善的生产体系和销售体系。城区成为以制瓷业为主的工商业集中场所，不仅是瓷器销售的集散地，也是瓷器生产的中心地。由此，产生了众多互相独立又彼此相依、各自不同分工的制瓷专门行业，及其辅助行业。行业中又有精细的专业分工，制瓷工艺日益复杂。宋应星《天工开物·陶埏》载："共计一杯工力，过手七十二方克成器，其中微细节目尚不能尽也。"

By the end of Qing Dynasty, Jingdezhen porcelain industry had formed a very perfect system of production and sales. The urban area became a centralized place for industry and commerce where was dominated by porcelain industry. It was not only a hub for porcelain sales, but also a center for porcelain production. As a result, many specialized trades of porcelain manufacturing with different division of labor appeared, which were both independent and dependent on each other, as well as their auxiliary industries. Labor division within this industry was gradually intensified, and the process of porcelain manufacturing became increasingly more complex. As noted in Taoshan (the chapter of ceramics), *Taoshan*

Tian Gong Kai Wu (Exploitation of the Works of Nature), "a piece of petuntse cannot be made into porcelain until it has gone through seventy-two procedures, which does not cover all the manufacturing details however."

自此，"七十二"成为景德镇制瓷繁复工序的概数。

Since then, "seventy-two" has become the approximate number of complicated processes of porcelain manufacturing in Jingdezhen.

清代宫廷画师孙祜、周鲲、丁观鹏以自己熟悉的山水为背景，配以景德镇的窑舍和劳作的工匠，绘制陶冶图 20 幅，记录官窑制瓷的详尽工艺。清高宗爱新觉罗·弘历之乾隆八年（1743 年），造办处将图交与景德镇督陶官唐英，命其按制瓷顺序编排，并撰写说明。当年，唐英以左图右文的形式编成《陶冶图说》。

During the Qing Dynasty, the court painters named Sun Hu, Zhou Kun and Ding Guanpeng drew 20 pictures of porcelain manufacturing, with their familiar landscapes as the background, together with kiln houses and working craftsmen in Jingdezhen, recording the detailed process of porcelain manufacturing at the official kilns. In the eighth year of the Qianlong reign (1743) of Emperor Gaozong (personal name Aixinjueluo Hongli), the Qing imperial workshops (Zaobanchu) handed over these pictures to Tang Ying, who was the superintendent of the imperial kiln factory in Jingdezhen, and ordered him arrange the pages in order according to the technical procedures and provide detail description of the technical processes. In that year, Tang Ying completed the compilation of *the Twenty Illustrations of the Manufacture of Porcelain* with pictures on the left side and texts on the right side.

七十二道工序可从《陶冶图说》中得到展现。

Seventy-two manufacturing processes of porcelain are elaborated in *the*

Twenty Illustrations of the Manufacture of Porcelain.

一为采石制泥。制瓷以瓷石做胎，瓷石产于安徽省祁门，山名坪里、谷口两处，距离窑厂二百里路。当地利用山间溪流设轮作碓，将瓷石春成泥，制成色纯质细的砖式原料，称为白不。

1. Mining for the stone and preparation of the paste

In the manufacture of porcelain, the body is formed of molded earth. The stone is found at the two mountains called Pingli and Gukou in Qimen, Anhui province, which is two hundred li distant from the porcelain manufactory. The natives take advantage of the mountain torrents to erect wheels provided with crushers. Having been finely pulverized, it is then purified by washing and levigation, and made up in the form of bricks, which are called "Baidun (petuntse)."

二为淘练泥土。制瓷所需要的瓷泥，需经淘练，使其精纯。瓷泥放入水缸浸泡、翻搅，使杂质下沉，再将泥浆舀到置于缸上的马尾细筛中过滤。过滤后的泥浆再分别注入过泥匣钵内沉淀，使泥浆稠厚成形。

采石制泥
Mining for the Stone and Preparation
of the Paste

淘练泥土
Washing and Purification of the Clay

2. Washing and purification of the clay

In the manufacture of porcelain, the first requisite is that of washing and purifying the materials of the paste so as to make it of fine homogeneous texture. The method of purifying the paste is to mix the materials with water in large earthenware jars and to stir the mixture so that it remains suspended in the water. While the impurities sink to the bottom, the paste is passed through a fine horsehair sieve. Then, it is poured into saggars, and the paste is left to become solidified.

三为炼灰配釉。配釉要用灰。御窑厂所用的灰出自景德镇之南一百四十里的乐平县。灰以青白石和凤尾草叠垒烧制而成，再配上练好的瓷泥，调成浆水。细泥与灰的比例按 10：1 调配，则为上品瓷用釉，以 7：3 或 8：2 的比例调配为中品之釉，对半或泥少灰多的则是粗釉。

3. Burning the ashes to formulate glazes

The composition used for glazing needs ashes. The ashes for the glaze for the imperial kilns come from Leping County, which is one hundred and forty li (70 km) to the south of Jingdezhen. They are made by burning a gray-colored limestone with ferns piled in alternate layers. The residue forms the ashes for the glaze. The finest kind of petuntse made into a paste with water is added to the liquid glaze ashes, and mixed to form a kind of seriflux. Ten measures of petuntse paste and one measure of ashes form the glaze for the highest class of porcelain. Seven or eight ladles of paste and two or three ladies of

炼灰配釉

Burning the Ashes to Formulate Glazes

ashes form the glaze for the middle class. If the paste and ashes are mixed in equal proportions, or if the ashes are more than the paste, the glaze is only fit for coarse wares.

四为制造匣钵。瓷坯入窑需要洁净，不得沾惹沙灰，所以需套入匣钵内烧制。制匣钵的泥土取自位于景德镇东北的里淳村和宝石山两地。原料不用淘洗，以轮车和拉坯法制作。

4. Manufacture of saggars

The porcelain while being fired in the furnace must be kept perfectly clean. A single spot of dirt makes a colored stain. For these reasons it is necessary to place the porcelain inside the saggars. The clay used in making these cases comes from Lichun village and Baoshi hill, which is on the northeast of Jingdezhen. The clay used in making these saggers does not need to be washed. The saggers are fashioned on a wheel, which is similar to the wheel used for porcelain.

制造匣钵
Manufacture of Saggars

圆器修模
Preparing Moulds for Round Wares

五为圆器修模。碗、盘、碟称作圆器。烧好后的成品因为窑火之故，会收缩，一尺之坯，得七八寸之器。所以，制造圆器，必须先有模子。模子是拉坯时的样品，要修改、试烧数次才能做好。

5.Preparing moulds for round wares

Bowls, plates and saucers are called round wares. The raw paste, which is expanded and loose in texture, becomes contracted and solidified to about seven or eight tenths of its original size during the process of firing. The proper proportionate size of the unbaked piece is fixed by the mold. Each piece must have several molds prepared, and the size and pattern of the contents when taken out of the kiln must be exactly alike.

六为圆器拉坯。圆器用轮车拉坯。就器之大小，又分为两种办法。大者拉一尺至两三尺之器，小者拉一尺以内之器。在一只大木盘下设轴即是轮车，另有工匠将练好的泥置于车盘上，拉坯工坐于车架上，以竹竿拨动木盘飞快转动，然后以手按泥，以目测定款式、大小，不差毫厘。

6. Throwing round wares on the wheel

Round wares are wheel thrown. In terms of their size, there are two processes of work in the manufacture of this round wares. The first is to take the large pieces from one up to two or three feet in diameter; the second is to make on the wheel the same kind of pieces which measure less than a foot across. The wheel

圆器拉坯

Throwing Round Wares on the Wheel

consists of a disk of wood mounted below upon a perpendicular axle. Beside the wheel is an attendant workman, who kneads the paste to a proper consistency and puts it on the table. The potter sits upon the border of the framework and turns the wheel with a bamboo staff. While the wheel is spinning round, he works the paste with both hands; it follows the hands, lengthening or shortening, contracting or widening, in a succession of shapes. It is in this way that the round ware is fashioned so that it varies not a hair's breadth in size.

七为琢器做坯。琢器指瓶、尊之类的器物。浑圆的琢器做法同于圆器，先拉坯，再用大羊毛笔蘸水洗磨光洁，然后或吹釉入窑即成白釉器，或绘画上釉即成青花瓷。八方、六方等棱角之器的制作方法，则是用布包泥以拍练成片，裁成块段，再用泥调糊粘合。

7. Fabrication of hollow wares

Hollow wares refer to vessels such as vases, zun that have a significant depth and volume. Plain round vases are fashioned upon the potter's wheel, in the same manner as ordinary round wares. After the vase has been shaped, it is washed with a large goat's-hair brush dipped in water, till the surface is perfectly bright and

spotlessly clean. After this, the glaze is blown on. Then, it is fired in the kiln, and comes out a piece of white porcelain. In making the octagonal, hexagonal and other angular vases, the paste, wrapped in cotton cloth, is pressed with flat boards into thin slabs, which are cut with a knife into sections. The pieces are joined together by some of the original paste diluted with water.

琢器做坯

Fabrication of Hollow Wares

八为采取青料。青料是绘青花和制作霁青釉的原料，出自浙江绍兴、金华两郡的山中。挖出的青料在溪流中洗去浮土，其色泽黑黄，大而圆者为上品，名为顶圆子。贩者携至烧瓷之所，埋入窑地锻炼三日，取出后淘洗售卖。

采取青料
Extracting Blue Pigment

8. Extracting blue pigment

The blue material is the raw material for painting blue and white and making deep blue monochrome glaze, which it is found in several mountains within the prefectures Shaoxin and Jinghua in Zhejiang province. The soil that adheres to the material is washed away in the stream after digged out. This kind of mineral is dark brown. The large round pieces that are called "best rounds" furnish the best blue. The material is brought by peddlers to the places for firing porcelain wares, which is buried by them under the floor of the furnace being roasted for three days. After taken out, it will be washed before ready for sale.

拣选青料
Processing and Purifying the Blue Pigment

九为拣选青料。青料炼出后，需要拣选，由料户专门负责。料分三等，黑丝润泽光色俱全者为上，用于仿古瓷、霁青釉、青花瓷。色虽黑绿而少润泽者，用于粗瓷。光色全无者不用。

9.Processing and purifying the blue pigment

The blue material, after it has been roasted, must be specially selected, so there is a particular class of workmen whose duty it is to attend to this. The blue material is divided into three grades. The selected superior kind is dark green with rich translucent tint and brilliant aspect, which is usually used in the imitation of antique porcelain, for the monochrome blue glaze, and for blue and white porcelain. The secondary class material in the same dark-green color, but wanting somewhat in richness and luster, is used for the decoration of the coarser porcelain made for sale. The inferior material, that has neither luster nor color, is picked out and thrown away.

印坯乳料

Moulding the Body and Grinding the Blue Pigment

十为印坯乳料。即印坯和乳料两个工艺。印坯：圆器拉好坯后，将修模套在坯上，以手拍按，使泥坯周正均匀，然后退下修模，阴干瓷坯以备旋削。其湿坯不宜日晒，否则会折裂。乳料：绘瓷所需的颜料，研乳宜细。每十两颜料为一钵专供研乳，研乳一月之后，才可应用。

10.Moulding the body and grinding the blue pigment

The process includes molding the body and grinding the blue pigment. After the large and small round pieces have been shaped on the wheel, they are put into the molds which have been previously prepared and are pressed gently with the hands until the paste becomes of regular form and uniform thickness. The piece is then taken out and dried in a shady place till it is ready to be shaped with the polishing knives. The damp paste must not be exposed to the sun, otherwise it will crack after heating. With regard to the preparation of the color for the artists, it must be ground perfectly fine in a mortar; Ten ounces of the material are put into each mortar, and a special class of workmen grinds it for a whole month before it is fit to be used.

圆器青花

Painting the Round Wares in Blue and White

十一为圆器青花。在圆器上绘青花，分工颇细。勾线、渲染，绘旋纹边饰（俗称打青箍）、花鸟禽鱼、人物、写款都各有专职，而且按类聚室操作，这是青花不同于五彩之处。

11. Painting the round wares in blue and white

The different kinds of round wares painted in blue and white are each numbered by the hundred and thousand. Those in charge of sketching the outlines, encircling blue bands, attaching the seals, painting flowers and birds, fishes and living objects each have their own specific job duty. In order to secure certain uniformity in their work, the sketchers and painters, although kept distinct, occupy the same house. The art of painting in blue and white differs widely from that of decoration in enamel colors.

十二为制画琢器。琢器的样式，有方、圆、棱角之别。制画的方法有彩绘、镂雕之异。仿旧须宗其典雅，肇新务审其渊源。制瓷之器，须遵古制。纹饰之作，花团锦簇，胜于春色。

12. Fabrication and decoration of hollow wares

Hollow wares include the square, the round, the ribbed, and those with prominent angles. There are various styles of decoration executed by painting in colors and carving in openwork. In copies from antiquity, artistic models must be followed; in novelty of invention, there is a deep spring to draw from. In the decoration of porcelain correct canons of art should be followed; the design should be taken from the patterns of old brocades and embroidery, the colors from a garden as seen in springtime from a pavilion.

制画琢器

Fabrication and Decoration of Hollow Wares

十三为蘸釉吹釉。上釉之法，古代制作长方、棱角的琢器用刷釉法，但不匀。大小圆器及浑圆琢器用蘸釉法，但又因体重而多破。今小圆器，仍于缸内蘸釉，其余的都用吹釉法。

13. Application of glaze by dipping and blowing

蘸釉吹釉

Application of Glaze by Dipping and Blowing

The ancient method of putting on the glaze was to apply it to the surface of the vase with a brush filled with the liquid glaze, whether it is square, tall, fluted, or ribbed, but it was difficult to distribute it evenly in this way. Round wares, both large and small, and plain round vases use the dipping method, but they might fail by being either too thickly or too thinly covered, and, besides, so many were broken that it was difficult to produce perfect specimens. Currently, small round pieces are still dipped into the large jar of glaze liquid, but others are glazed by blowing.

旋坯挖足

Scraping the Body and Cutting the Foot

十四为旋坯挖足。即旋坯和挖足两道工序。旋坯：圆器尺寸由模子而定，但光平度要通过旋坯修整。旋坯车，形似拉坯车，不同的是在木盘中向上立一根木桩，顶上包垫丝锦以护坯。将坯合扣于桩上，拨盘转旋，以工匠持刀旋削，则器之里外皆得平光，是技术性很强的工种。挖足：在拉坯时，需留二三寸长的泥靶，以便于把握好画坯、吹釉。等吹画完工，即旋去泥靶，挖足写款。

14. Scraping the body and cutting the foot

There are two processes: scraping the body and cutting the foot. In the process scraping the body, the size of the round piece has been fixed in the mold, but the smooth polish of the surface depends on the polisher, whose province is another branch of work, that of "turning." The polishing wheel is like the ordinary potter's wheel; the difference is that it has projecting upward in the middle a wooden mandrel. The top of this mandrel, which is rounded, is wrapped in raw silk to protect the interior of the piece from injury. The piece about to be turned is put upon the mandrel. When the wheel is spun round, it is pared with the knife till both the inside and outside is given the same perfectly smooth polish. The coarser or finer finish of the form depends upon the inferior or superior handiwork of the polisher, whose work is consequently of great importance. With regard to the next process-cutting the foot, when first fashioned upon the potter's wheel, it is necessary to have a paste handle left under the foot two or three inches long, by which it is held while it is being painted and glazed. After the glazing and the painting of the decoration are finished, this handle is removed by the polisher, who at the same time scoops out the foot, which the mark afterwards is written underneath.

十五为成坯入窑。窑作长圆形如覆瓮，高宽丈许，上罩大瓦屋，名窑棚。窑棚后部之外，设烟囱，高二丈余。烧窑有专门的窑户。瓷坯做好后，装在匣钵里，送至窑户。待满窑后，开始发火，随即将窑门以砖砌好，只留一方孔投松柴，片刻不停。待窑内匣钵呈现银红色后止火，再等一昼夜后开窑。

15. Stacking the finished wares into the kiln

The kiln is long and round, and resembles in shape a large water jar (weng) turned over on its side. It measures a little over ten feet in height and

成坯入窑

Stacking the Finished Wares into the Kiln

breadth, about twice as much in depth. It is covered with a large, tiled building which is called the "kiln-shed." The chimney rises to a height of over twenty feet behind, outside the kiln-shed. The porcelain, when finished, is packed in the saggars and sent out to the furnace men. When these men put it in the kiln, they arrange the saggars in piles, one above the other, in separate rows, so as to leave a space between the rows for the free passage of the flames. After the kiln has been fully charged, the fire is lighted, and the entrance is then bricked up, leaving only a square hole, through which billets of pinewood are thrown in without intermission. When the saggars inside the furnace have attained a silvery red color (white heat), the firing is stopped, and after the lapse of another twenty-four hours, the kiln is opened.

烧坯开窑

Opening the Kiln When the Firing Is Finished

十六为烧坯开窑。烧制瓷器，依赖于窑火。从入窑到开窑大约三日，到第四日清晨开窑。窑中套装瓷器的匣钵尚带紫红色，人不能近，唯开窑之匠用数十层布制成手套蘸以冷水护手，并用湿布包裹头面肩背，方能入窑搬取瓷器。

16. Opening the kiln when the firing is finished

The perfection of the porcelain depends upon the kiln firing, which, reckoning from the time of putting in to that of taking out, usually occupies about three days. On the fourth day, the furnace is opened early in the morning. The saggars inside it, which contain the porcelain, are still of a dull-red color, so it is impossible to enter yet. After a time, the workmen who open the kiln, with their hands protected by gloves of ten or more folds of cotton soaked in cold water, and with damp cloths wrapped around their heads, shoulders, and backs, are able to go in to take out the porcelain.

　　十七为圆琢洋彩。在圆器和琢器上模仿西洋的技法绘彩画，所以称为洋彩。要选画瓷高手，将原料调细，先用瓷片画染试烧，以熟悉颜料的火性，才能由粗及细，熟中生巧。画匠以眼明心细者为佳。粉彩瓷所用颜料与珐琅色相同。其调色之法有芸香油、胶水、清水三种。

圆琢洋彩

Round Ware and Vases Decorated in Foreign Style

　　17. Round ware and vases decorated in foreign style

　　Both round wares and hollow wares are painted in enamel colors by imitating western style, which is consequently called Yangcai, or foreign colors. Artisans with superior skills are selected to paint the decoration. To begin with, the different materials of the color have been previously finely ground and properly combined. The artisan first paints with them upon a slab of white porcelain, which is fired to test the properties of the colors and the length of firing they require. He is gradually promoted from coarse work to fine, and acquires skill by constant practice, which needs a good eye, attentive mind, and exact hand being required to at tain excellence. The colors which are employed for famille rose are the same as those used for the enamel color. They are mixed with three different kinds of medium, i.e., rue oil, liquid glue, and pure water.

明炉暗炉
The Open Stove and the Close Stove

十八为明炉暗炉。制作釉上彩瓷，需先在窑内烧成白釉瓷，而后于上绘纹饰，再于低温明炉或暗炉内烘烤以固定颜色。小件用明炉。明炉周围有炭火，炉内置铁轮，炉下托以铁叉，将瓷器送入炉内，以铁钩拨轮令其旋转，以匀火气。大件则用暗炉。炉高三尺，径二尺六七寸，周围夹层以贮炭火，下留风眼。将瓷器贮于膛内，炉顶盖板用黄泥封固，工匠手持圆板以避火气，烧一昼夜为度。

18. The open stove and the close stove

White paste porcelain that has been previously fired in the furnace is first decorated by the artisan with painting in colors. When it has been painted in colors, it must be again fired to fix the colors. For this purpose, two kinds of muffle stoves are used with one kind being open and the other closed. The open stove is used for the smaller pieces. When the charcoal fire has been lighted inside, the porcelain is placed upon an iron wheel. The wheel is supported upon an iron fork, by which the porcelain is passed into the stove. The fireman holds an iron hook with his other hand so that he may be able to turn the wheel around in the fire to equalize the action of the heat. For large pieces, the closed stove is employed. This stove is three feet high and nearly two feet and three-quarters in diameter with a double wall to hold the charcoal fire, and the wall, being perforated below for the entrance of air, surrounds it. The porcelain is introduced into the interior of the stove, while the artisan holds a circular shield to protect him from the heat of the fire. The top of the stove is then closed by a flat cover of yellow clay and closely luted. In general, the firing takes a period of about twenty-four hours.

十九为束草装桶。瓷器烧成后按上色、二色、三色、脚货四类分拣。后两档于本地销售。上色圆器及上色、二色琢器用纸包装桶。二色圆器每十件用草包扎，每十件为一桶，以便运载。各省所用的粗瓷只用菱草和竹篾包扎。

19.Wrapping in straw and packing in cases

After the porcelain has been taken out of the furnace, it is arranged into four separate classes, which are

束草装桶
Wrapping in Straw and Packing in Cases

known by the names of first-class color, second-class color, third-class color, and inferior ware. Porcelain wares of third-class color and inferior wares are kept back for local sale. Round wares of first-class color and hollow wares of the first and second class are all wrapped up in paper and packed in round cases. Round wares of second-class color are tied together in bundles, each composed of ten pieces, for convenience of carriage to distant areas. Coarser porcelain wares, distributed

throughout the different provinces, are only tied up in bundles with reeds and mat
ting.

祀神酬愿

Worshipping the God and Offering Sacrifice

二十为祀神酬愿。景德镇方圆十余里，山环水绕，中央一洲。缘于产
瓷之地，商贩聚集。民窑二三百区，终岁烟火相望。工匠数十万，皆以烧
瓷为生，故窑火之事，皆尚祈祷祭祀。

20.Worshipping the god and offering sacrifice

Jingdezhen has a circumference of over ten li surrounded by rivers and
mountains with an islet in the center. On account of its porcelain production, merchants throng to it from far and near. About 200 or 300 private kilns, exhibit a
constant succession of flames and smoke the whole year round, and give employment to not less than several hundreds of thousands of potters and artisans. The
porcelain industry gives subsistence to an immense number of people whose life
hangs on the success or failure of the furnace fires, so they are all devout in worship and sacrifice.

明御器厂

The Imperial Ware Factory of the Ming Dynasties

明代十六帝，历时 276 年。

The Ming Dynasty, which encompassed the reigns of 16 emperors, lasted 276 years.

太祖朱元璋之洪武二年（1369 年），朝廷延续元代官窑制度，在浮梁瓷局的基础上扩建，设厂制陶，以供尚方之用。宣宗朱瞻基即位之初的宣德元年（1426 年），在饶州府治所在地鄱阳县城内月波门外，设立景德镇御器厂官衙，为每年景德镇进贡瓷器时太监检验贡瓷器质量的场所。至此，专

景德镇御窑厂

ngdezhen Imperial Kiln Factory

供御用成为景德镇窑场的最大特点，由此完成了由传统官窑向御窑的转变。

In the second year of the Hongwu reign of Emperor Taizu (personal name Zhu Yuanzhang) of the Ming Dynasty (1369), the imperial court continued to carry on the official kiln system of the Yuan Dynasty, building a factory based on the Fuliang Porcelain Bureau, "The factory makes porcelain wares for the Shangfang (An official office for making artifacts used by emperors in ancient times)." In the first year of the Xuande reign (1426), when Zhu Zhanji ascended the throne to become Emperor Xuanzong, an official office of Jingdezhen imperial ware factory was established outside Yuebo gate in Poyang County, where Raozhou Prefecture was located. It is a place for eunuchs to inspect the quality of Jingdezhen tribute porcelain every year. So far, the official kilns in Jingdezhen were commissioned to produce porcelain exclusively for the imperial court, which had completed the transformation from traditional official kiln to the imperial kiln.

明代官窑窑场在景德镇珠山之南，周长五里许，面积约为5.43万平方米。

The official kiln workshop of the Ming Dynasty was located in the south of Zhushan mountain in Jingdezhen, with a circumference of about 2.5 kilometers and an area of about 54,300 square meters.

御窑厂遗址
The Relic Site of the Imperial Kiln Factory

自洪武官窑设立，历建文、永乐、洪熙、宣德、正统、景泰、天顺、成化、弘治、正德、嘉靖、隆庆、万历各朝，约 240 年。武宗朱厚照之正德初年（约 1506 年），恢复因战事停烧的御器厂。神宗朱翊钧之万历三十六年（1608 年），窑厂停烧。

Since the establishment of the Hongwu official kiln, it had lasted for about 240 years throughout different dynasties, including Jianwen, Yongle, Hong Xi, Xuande, Zhengtong, Jingtai, Tianshun, Chenghua, Hongzhi, Zhengde, Jiajing, Longqing and Wanli. In the early year of the Zhengde reign　(about1506) of Emperor Wuzong　(personal namc Zhu IIouzhao), the imperial ware factory resumed production, which had stopped operating due to war. In the thirty-sixth year of the Wanli reign (1608) of Emperor Shenzong (personal name Zhu Yijun), the imperial kiln factory stopped operation.

景德镇御器厂为明朝历代皇帝烧造了无数精美的瓷器。以下择要介绍。

Jingdezhen imperial ware factory fired countless exquisite porcelain for the emperors of the Ming Dynasty, which is briefly summarized as follows.

太祖朱元璋之洪武年间（1368～1398 年），陶厂始创。凭借皇家特权，景德镇汇聚了全国瓷艺资源，极大地推动了瓷器工艺的革新。明以红为贵为吉，洪武盛行红釉瓷器，器型硕大的青花釉里红大盘、大碗、大罐，素朴粗豪，形成洪武御窑气度雄浑的独特风格。

The imperial porcelain factory was founded during the Hongwu reign period of Emperor Taizu　(personal name Zhu Yuanzhang) of the Ming Dynasty (1368-1398). With the royal privilege, Jingdezhen gathered the national porcelain art resources, which greatly promoted the innovation of porcelain making techniques. In the Ming Dynasty, red was regarded as fortune and auspiciousness, so red glazed porcelain was popular during the Hongwu reign period. Blue and white underglaze red plates, bowls and jars in large size were plain and vigorous, forming the unique style of porcelain from the Hongwu imperial kilns.

明御器厂出土永乐白釉三壶连通器
A Three Conjoined Vase with Clear Glaze of the Yongle Era
Unearthed in the Ming Imperial Ware Factory

　　成祖朱棣之永乐年间（1403～1424 年），御器在红色之中，添加了白色，最为名贵的瓷器有三种：甜白釉瓷器为常，青花瓷器为贵，鲜红釉瓷器为宝，号称"永乐三宝"。永乐开始以年号作款，延续至明清两代。

　　During the Yongle reign of Emperor Chenzhu (personal name Zhu Di) of the Ming Dynasty （1403-1424）, white color was decorated on red ground in the process of manufacturing imperial wares, and three varieties of the most precious porcelain, known as the "Three Treasures of Yongle", were created, among which sweet white glazed porcelain was considered as favoured objects, blue and white porcelain as precious objects, and bright red glazed porcelain as treasured objects. The practice of painting reign marks on porcelain was established during the Yongle reign, which had been preserved to the Ming and Qing Dynasties.

宣宗朱瞻基之宣德年间（1426～1435 年），官窑以产品量多且质优，被称为历代官窑之冠。《景德镇陶录》说："宣窑器无物不佳，小巧尤妙，此明窑极盛时也。"宣德时期景德镇御器厂烧造的青花蟋蟀罐最负盛名，祭红器也是重要品种。

During the Xuande reign (1426-1435) of Emperor Xuanzong (personal name Zhu Zhanji), the official kiln was known as the crown of the official kilns of all dynasties for its large quantity and high quality of products. According to *Records of Jingdezhen Ceramics*, "Xuanyao porcelain wares were featured with small and exquisite size and high quality with very few exceptions, which was in its heyday of the Ming kiln." During the Xuande period, the blue and white cricket pots made by Jingdezhen imperial ware factory were the most famous, and "Sacrificial Red Porcelain" (Jihong) was also an important variety.

明宣德青花折枝花卉纹八方烛台
A Blue and White Octagonal Candlestick with Floral Sprays, Xuande Period, Ming Dynasty

永乐和宣德时期的青花瓷器，胎釉精细、青色浓艳、造型多样、纹饰优美，被称为中国青花瓷器的黄金时代。宣德青花与永乐青花并称为永宣青花，二者相映成趣，形成一种别有韵味的色调。

Yongle and Xuande period was known as the golden age of Chinese blue and white porcelain because of its fine body glaze, rich blue color, diverse shapes and beautiful decorative patterns. Both Xuande blue and white and Yongle blue and white were widely recognized as "Yongxuan blue and white porcelain" that complemented each other, creating a distinctive tone.

宣德朝之后的正统、景泰、天顺三朝（1436～1464年），因内忧外患、经济萧条、民生凋敝，景德镇瓷业受到影响，绝大多数官窑没有纪年款，一度被称作空白期。但综合文献记载、景德镇出土遗物、传世品等情况可知，此时期并非完全空白，官窑烧造仍有一定规模，产品也有较高的水平。

During the Zhengtong, Jingtai and Tianshun reign periods (1436-1464) after the Xuande reign, Jingdezhen porcelain industry was greatly affected by domestic turmoil and foreign aggression; therefore, vast majority of porcelains from the official kilns did not bear reign marks, which was once called "the blank period." However, according to the comprehensive analysis of literature records, unearthed relics and handed down vessels in Jingdezhen, this period is not completely blank. The porcelain production at the official kilns still maintains a certain scale, and

明成化斗彩鸡缸杯
A Doucai (joined colours) Chicken Cup, Chenghua Period, Ming Dynasty

the quality of products is high.

宪宗朱见深之成化年间（1465～1487年），官窑瓷器已改变了永乐和宣德以来雄健豪放的风貌，创烧了斗彩，瓷器造型玲珑秀气，胎体细润晶莹，彩料精选纯正，色调柔和宁静，绘画淡雅幽婉，斗彩鸡缸杯为标志性器物。

During the Chenghua reign (1465-1487) of Emperor Xianzong (personal name Zhu Jiansheng), the imperial kiln porcelain had changed the vigorous style inherited from the Yongle and Xuande reign. Doucai (contrasting colors) was fired. Stylistically, porcelain pieces were celebrated for exquisite and delicate shape, fine and crystal body, a selection of pigments with pure color, soft and quiet tone, and elegant and gentle painting. Among them, doucai chicken cup was a representative one.

孝宗朱祐樘之弘治（1488～1505年）和武宗朱厚照之正德（1506～1521年）时期，官窑烧造少，品种和器型不多。弘治瓷器装饰纹样沿袭成化，缺少变化。娇黄的黄釉瓷器为其极品。正德瓷器品种和器型略比弘治丰富，八思巴文和阿拉伯文装饰最具特色。

During the Hongzhi reign (1488-1505) of Emperor Xiaozong (personal name Zhu Youcheng) and the Zhengde reign (1506-1521)of Emperor Wuzong (personal name Zhu Houzhao), the yield, quality, variety and shape of porcelain all shrunk rapidly. In terms of decorative patterns, porcelain pieces of the Hongzhi reign were thought to have continued in the same way as those of the Chenghua reign, lack of change. Porcelain pieces with the delicate yellow glaze were the best. Porcelain pieces of the Zhengde reign were slightly richer in varieties and types than those of the Hongzhi reign, and porcelain pieces decorated with the Phags-pa script and Arabic script were the most distinctive.

万历青花五彩龙纹碗
A Wucai Blue and White Bowl with Dragons, Wanli
Period, Ming Dynasty

世宗朱厚熜之嘉靖（1522～1566 年）、穆宗朱载垕之隆庆（1567～
1572 年）和神宗朱翊钧之万历（1573～1620 年）时期，瓷器工艺进一步
得到发展，产量很大。嘉靖时期，崇奉道教，多产大器，流行葫芦瓶，器
物常采用八卦、云鹤、璎纹和卍字纹装饰。隆庆时期，开放海禁，瓷器远
销海外，瓷业兴旺。万历时期，以五彩瓷为代表，多大面积敷彩，莲瓣逐
渐图案化，常用婴戏题材，也制作大器。

During the Jiajing reign （1522-1566) of Emperor Shizong （personal name
Zhu Houcong) and the Longqing reign （1567-1572) of Emperor Mu Zong（per-
sonal name Zhu Zaihou) and the Wanli reign （1573-1620) of Emperor Shenzong
(personal name Zhu Yijun), porcelain making techniques were further developed
and the output was extremely large. During the Jiajing reign, owing to respect for
Taoism among people, large-size utensils were produced, and gourd bottles were
popular. The utensils were often decorated with eight diagrams, cloud cranes,
Yingluo patterns and 卍 pattern patterns. During the Longqing reign, with the sea
ban dismissed, porcelain wares were exported overseas, so porcelain business was
extremely booming. Wanli period porcelain, which was represented by poly

chrome porcelain, was generally decorated with a large color area, and lotus flower petals decoration was gradually patterned. In addition, the decorative pattern of "children-at-play" was frequently used, and large vessels were also made.

自嘉靖时期开始，施行官搭民烧制度，官窑器与民窑器差别缩小。

Since the Jiajing Dynasty, the system of "moulding by imperial kiln and firing by civil kiln" was implemented, and the differences between official kiln wares and civilian kiln wares were narrowed.

清御窑厂

The Imperial Kiln Factory of the Qing Dynasty

清建国于 1616 年，初称后金，1636 年始改国号为清，1644 年入关，历时 296 年。自顺治帝迁都北京，共十帝，268 年。

Early declared as the Later Jin Dynasty, the Qing Dynasty was established in 1616, changing its name to "Qing" in 1636, and became the ruler of all of China in 1644, which lasted 296 years. Since Emperor Shunzhi moved its capital to Beijing city, having altogether 10 emperors for a period of 268 years.

清代延续明代官窑制度，在景德镇珠山设置御窑厂。《景德镇陶录·国朝御窑厂恭记》记载，清代"建厂造陶"，始于顺治十一年（1654 年）。《浮梁县

景德镇御窑厂国家考古遗址公园
The Imperial Porcelain Factory National Archaeological Site Park

志·陶书》也记载，这一年，景德镇"奉旨烧造龙缸"。圣祖爱新觉罗·玄烨之康熙十九年（1680 年）以后，时局稳定，臧应选到景德镇督陶，景德镇御窑厂开始正常生产。

The Qing Dynasty continued the official kiln system of the Ming Dynasty and set up an imperial kiln factory at Zhushan in Jingdezhen. According to *Records of Jingdezhen Ceramics*, the imperial kiln factory started shaping and firing porcelain objects in the 11th year of Shunzhi of the Qing Dynasty （1654). As recorded in Taoshu, *Fuliang County Chronicle*, "dragon vats were produced by order of the emperor at the same year" in Jingdezhen. During the Shunzhi reign period of Emperor Shizu （personal name Aixinjueluo Fulin) of the Qing Dynasty (1644-1661), the situation was stable. Because the political situation has been rel-

atively stable since 1680, the 19th year of the Kangxi reign of Emperor Shengzu (personal name Aixinjuero Xuanye), Zang Yingxuan was assigned to supervise porcelain production in Jingdezhen, and then Jingdezhen imperial kiln factory began to resume normal production.

清代镇窑

Jingdezhen Kiln of the Qing Dynasty

清御窑厂的规模与明御器厂大致相当。

The scale of the Qing imperial kiln factory is roughly the same as that of the Ming imperial ware factory.

清代，基本沿袭明代御器厂的厂署规制，但进行了三项重大改革。一是改强迫性的匠籍制为相对自由的雇募制。二是对垄断的制瓷原料相对地解除了禁令。三是督陶官由素质较高的朝廷官员担任，避免了太监专权的影响。

During the Qing Dynasty, the regulation of the imperial ware factory of the Ming Dynasty was basically followed, but three major reforms were implemented. Firstly, the forced craftsmanship system was changed into a relatively free employment and recruitment system. Secondly, the ban on monopolized porcelain raw materials was relatively removed. Thirdly, pottery supervisors were appointed by high qualified and experienced imperial officials, which avoided the deleterious impact of eunuchs' dictatorship.

康、雍、乾三朝，景德镇御窑瓷器生产臻于巅峰，达到历史最高水平。

During the three reigns of Emperor Kangxi, Yongzheng, and Qianlong, the manufacture of porcelain at Jingdezhen imperial kiln factory reached its peak of development: seeing historic heights.

清康熙胭脂红地珐琅彩开光花卉纹碗
A Falangcai (foreign colours) Bowl with Floral Panels on a Rouge-red Ground, Kangxi period, Qing Dynasty

圣祖爱新觉罗·玄烨之康熙年间（1662～1722年），御窑厂瓷器品种丰富，大致可分为釉下彩瓷、釉上彩瓷、釉下彩与釉上彩相结合的彩瓷、颜色釉瓷和杂釉彩瓷、素三彩瓷等。其中，创烧珐琅彩瓷，郎廷极督造出郎红釉，青花瓷被推为清代青花之冠，还烧制了大量笔筒、水洗、臂搁、瓷砚等文房用瓷。

During the Kangxi reign (1662-1722) of Emperor Shengzu (personal name Aisinjuero Xuanye), the variety of porcelain objects produced in the imperial kiln factory was greatly extended, which could be roughly divided into underglaze color porcelain, overglaze color porcelain, color porcelain combined with underglaze color, color glazed porcelain and multi-colored glaze porcelain, plain tricolor porcelain and so on. Among them, enamel colored porcelain was created; Lang kiln red glaze was developed under the supervision of Lang Tingji; blue and white porcelain was promoted as the crown of blue and white porcelain in the Qing Dynasty. In addition, a large number of stationery porcelain pieces were produced,

清雍正粉彩三果纹碗
A Famille-rose Bowl with Three Fruits Pattern, Yongzheng Period, Qing Dynasty

including pen pot, brush washer, wrist rest, porcelain inkstone, etc.

世宗爱新觉罗·胤禛之雍正年间（1723～1735 年），追求雅致恬淡，珐琅彩瓷兴盛，粉彩最有成就，瓷器精莹纯全，胎质白润细腻。年希尧督陶，成效卓著。御窑厂仿烧前朝作品达到高潮，尤以仿烧宋代五大名窑的色釉及明代永乐、宣德、成化三朝的青花最具水准。

During the Yongzheng reign (1723-1735) of Emperor Shizong (personal name Aixinjueluo Yinzhen), tranquility and elegance had been the artistic realm of porcelain manufacturing. With enamel colored porcelain flourished, famille rose won the highest acclaim. Porcelain was characterized by pure and translucent with white and delicate porcelain body. Nian Xiyao, a famous superintendent made an extraordinary contribution to porcelain development. The imitation firing of the works of the former dynasties at the imperial kilns reached its zenith. Especially objects by imitating the colored glaze porcelain of the five famous kilns of the Song Dynasty and blue and white porcelain of the Yongle, Xuande and Chenghua reigns of the Ming Dynasty were the most accomplished.

高宗爱新觉罗·弘历之乾隆年间（1736～1795 年），是中国瓷器集大

清乾隆粉彩九桃瓶

A Famille-rose Vase with Nine Peaches Design, Qianlong Period, Qing Dynasty

成时期，景德镇制瓷达到鼎盛，也是中国传统制瓷业由盛而衰的转折时期。唐英督陶，既仿古集成，又采今创新。粉彩装饰工艺渐趋繁缛，珐琅彩瓷趋于极盛，用于陈设的转心瓶、夹心套瓶、交泰瓶等高难度瓷器是为极品。

The Qianlong reign (1736-1795) of Emperor Gaozong (personal name Aix-injueluo Hongli) was the period of the integration of Chinese porcelain. Jingdezhen porcelain reached its zenith, which was also the turning point of the decline of Chinese traditional porcelain industry. When Tang Ying supervised porcelain production in Jingdezhen, he did not simply imitate ancient ways of making porcelain, but pursued technological innovation. Famille rose porcelain began to become elaborate, while the enamel colored porcelain gradually reached a climax of perfection. Those porcelain wares with difficult manufacturing process used for furnishings, including vases with revolving inner, vases with filling and Jiaotai vases and so on, were the best.

历史规律，盛极必衰。康雍乾之后，御窑瓷器一代不如一代。

According to the law of history, things start to decline after they reach the extremity of prosperity. After the Kangxi, Yongzheng and Qianlong reigns, porcelains from the imperial kiln factory were inferior from generation to generation.

仁宗爱新觉罗·颙琰之嘉庆年间（1796~1820 年），景德镇制瓷已明显不如乾隆时期，品种大大减少，风格基本与乾隆时期的瓷器相似，创新之作极少，留存下来的嘉庆官窑传世品数量相对减少，而且制作质量也不十分高。

During the Jiaqing reign (1796-1820) of Emperor Renzong (personal name Aixinjueluo Yongyan), the manufacture of porcelain in Jingdezhen was obviously inferior to that of the Qianlong reign, and types of objects were plummeting. The style of various types was basically similar to that of the Qianlong reign, and there were few innovative objects. The number of handed down products of Jiaqing official kilns remained was relatively reduced, and the production quality was not very high.

清道光珊瑚红地白梅花纹盖碗
A Box-bowl and Cover with Coral-red Glaze and White Plum
Pattern, Daoguang Period, Qing Dynasty

宣宗爱新觉罗·旻宁之道光年间（1821～1850 年），御窑瓷器在造型、釉彩和制作方面，有不尽如人意的草率作风，但同前朝嘉庆和以后几朝相比较，仍然有出色之处，不少产品远胜过嘉庆时期。

During the Daoguang reign (1821-1850) of Emperor Xuanzong (personal name Aixinjueluo Minning), although porcelain pieces made in the imperial kilns were quite unsatisfactory for the hasty style in modeling, glazing and manufacturing, they were still quite excellent compared with those of the previous Jiaqing reign and the following reigns, among which many products were far better than those of the Jiaqing reign.

文宗爱新觉罗·奕詝之咸丰年间（1851～1861 年），初年，太平军在江西广泛活动，景德镇御窑厂烧造的瓷器很难运出江西。咸丰五年（1855 年），景德镇御窑厂在战乱中停烧，传世的咸丰景德镇御窑瓷应为此前烧造。

In the first year of the Xianfeng reign (1851-1861) of Emperor Wenzong (personal name Aixinjueluo Yizhu), Jiangxi was frequently harassed by Taiping troops. Therefore, it was difficult to transport the porcelain objects fired by Jingdezhen imperial kilns out of Jiangxi. In the fifth year of the Xianfeng reign (1855), the imperial kiln factory in Jingdezhen ceased operation during the war. The handed down porcelain pieces from the imperial kilns of the Xianfeng period should be made before the war.

穆宗爱新觉罗·载淳之同治年间（1862～1874 年），经历了咸丰时期激烈的战乱之后，同治五年（1866 年）景德镇御窑厂恢复烧造，但由于连年战乱，窑业元气大伤，御窑厂烧造仅可勉强应付朝廷差事。

During the Tongzhi reign (1862-1874) of Emperor Muzong (personal name Aixinjueluo Zaichun), the imperial kiln factory in Jingdezhen resumed operation in the fifth year of the Tongzhi reign (1866) after the chaos caused by fierce war during the Xianfeng period. However, due to years of war, the vitality of the kiln industry was seriously damaged, so the imperial kiln factory could only barely meet the needs of the imperial court.

清光绪大雅斋款绿地粉彩花鸟纹高足盘
A Famille-rose Stem Plate with Birds and Flowers on a Green Ground and Mark of "Studio of the Great Elegance" (Daya Zhai), Guangxu Period, Qing Dynasty

德宗爱新觉罗·载湉之光绪年间（1875～1908年），御窑厂烧造的瓷器数量很多，传世品及品种也极为丰富，有落慈禧寝宫储秀宫和大雅斋款的产品。与前朝相似，仍以粉彩和青花为主流。制作质量比同治官窑器略胜一筹。

During the Guangxi reign (1875-1908) of Emperor Dezong (personal name Aixinjueluo Zaitian), porcelains fired at the imperial kiln factory were not only in great quantity, but also in great variety, among them there were some handed-down artifacts, including the porcelain wares exclusively used by the Empress Dowager Cixi at the Palace of Accumulated Elegance (Chuxiu Gong) and Dayazhai-Style wares. Similar to the previous dynasty, it was still dominated by famille rose (fencai) porcelain and blue and white porcelain. Its production quality was slightly better than that of Tongzhi official kilns.

爱新觉罗·溥仪之宣统年间（1909～1911年），御窑厂无特别创新，烧造量有限，品种不多，主要有青花、五彩、粉彩及各颜色釉等。浅绛粉彩在光绪时期流行后，此时仍继续发展，还烧造了停烧已久的珐琅彩瓷。

During the Xuantong reign (1909-1911) of Emperor Xuantong (personal name Aixinjueluo Pu Yi), the imperial kiln factory did little in the innovation of products with limited production quantity and few varieties, mainly including blue and white porcelain, polychrome porcelain, famille rose porcelain and various other color-glazed porcelain, etc. Since Qianjiang (a variant of fencai) became popular during the Guangxu reign, it had still been continuing to develop during this period. Moreover, enamel colored porcelain, which had ceased firing for a long time, resumed its production.

清宣统三年（1911年），辛亥革命成功，彻底推翻了帝制。景德镇御窑厂也寿终正寝，结束了自明洪武二年（1369年）以来542年皇家制瓷工厂的历史。

In the third year of the Xuantong reign of the Qing dynasty (1911), the revolution of 1911 succeeded and completely overthrew the monarchy. Jingdezhen imperial kiln factory also ceased operation, ending the 542-year history of imperial kiln porcelain factory since the second year of the Hongwu reign of the Ming Dynasty (1369).

粉彩花卉过墙枝大盘

A Large Famille-rose Plate with Design of Flowers on Branches

清朝覆亡、御窑厂撤销之后，袁世凯称帝，又改设陶务监督署，烧造过一批瓷器，世称洪宪瓷。袁世凯称帝幻灭后，陶务监督署也随即撤销。

After the fall of the Qing Dynasty and the revocation of the imperial kiln factory, Yuan Shikai declared himself emperor, and set up a porcelain supervision office. A batch of porcelain objects were fired in Jingdezhen, which were later referred to as "Hongxian Porcelain." After Yuan Shikai was forced to abdicate as emperor, the porcelain supervision office was immediately dissolved.

昔日御窑厂围墙
The Wall of the Former Imperial Kiln Factory

民国时期，在御窑厂西北侧，曾设立官助民办的江西瓷业公司，部分厂舍被军警屯驻，厂内许多建筑变成了断瓦颓垣。中华人民共和国成立之时，景德镇御窑厂内只剩下一座龙珠阁。

During the period of the Republic of China, Jiangxi Porcelain Company, a semi-government company funded by private owners was once established in the northwest of the imperial kiln factory, where some factory buildings were stationed by the security forces, and many buildings in the factory fell into broken bricks and tiles. When the People's Republic of China was founded, there was only Longzhu Pavilion left at Jingdezhen Imperial Kiln Factory.

器走天下

Sending Porcelain to All Parts of the World

《景德镇陶录·陶说杂编》说："昌南镇陶器，行于九域，施及外洋。"

As mentioned in "Miscellaneous collections of Taoshuo" in *Records of Jingdezhen Ceramics*, "Jingdezhen alone has the honor of sending porcelain to all parts of the world."

明成祖朱棣之永乐年间（1403～1424 年），大明皇帝命太监郑和率领一支 317 艘船、27800 多人的舰队出海，史称"郑和下西洋"。从永乐三年（1405 年）到宣宗朱瞻基之宣德八年（1433 年），一共出行 7 次，到达西南太平洋、南亚、印度洋、东非等地，历经 30 多个国家和地区，最远到达红海、非洲东海岸的索马里和肯尼亚。其中的一次出航，景德镇就奉旨烧造了 44.35 万件瓷器。

During the Yongle reign of Emperor Chenzhu (personal name Zhu Di) of the Ming Dynasty (1403-1424), the Empire of the great Ming ordered Zheng He, a eunuch, to lead a fleet of 317 ships and 27,800 people across the seas, which was known as "Zheng He's voyage to the west seas." Between the third year of the Yongle reign period (1405) and the eighth year of the Xuande reign period (1433), Zheng He made at least seven major excursions, traversing the South Chi-

明代葫芦窑

The Gourd Shaped Kiln of the Ming Dynasty

na Sea, South Asia, the Indian Ocean and East Africa, and visited over 30 countries and regions. On these trips, Zhen He's farthest voyage reached to the Red Sea and the coast of East Africa, stopping in today's Somalia and Kenya. On one of the voyages, it was said that 4,435,000 pieces of porcelain wares were fired in Jingdezhen by imperial decree.

外销青花盖盆
A Chinese Export Blue and White Tureen

随着 16 世纪大航海时代来临，东西方各条新航线不断开辟，景德镇外销瓷也进入全球化的时代。明中晚期至清初 200 余年，是景德镇瓷器外销的黄金时期。景德镇青花瓷成为世界瓷器市场上的最主要商品。从万历时期（1573~1620 年）开始，销往欧洲的瓷器几乎全是青花瓷。

With the advent of the great navigation era in the 16th century, new routes between the East and the West have been continuously opened up; therefore, Jingdezhen export porcelain has also entered the era of globalization. More than 200 years from the middle and late Ming Dynasty to the early Qing Dynasty was the golden period for Jingdezhen export porcelain. Jingdezhen blue and white porcelain has become the most important commodity in the world market for porcelain. Started from the Wanli reign of the Ming Dynasty(1573-1620), almost all export porcelain was of the blue and white variety.

纹章瓷壶

A Tea Pot with Coat of Arms

　　中国外销瓷是第一个扮演世界性商品的产品。外国学者估算，单是从明代末期到清代中期，由欧美公司组织运输和销售的中国瓷器，应有3亿多件。明代嘉靖后期至清代中期，景德镇生产的专供欧洲和美洲市场的外销瓷器是中国外销瓷的主要代表。

　　Chinese export porcelain is truly the world's first global commodity. Foreign scholars estimate that from the late Ming Dynasty to the middle Qing Dynasty alone, there should be more than 300 million pieces of Chinese porcelain transported and sold by European and American companies. During the period from the Jiajing reign of the Ming Dynasty to the middle of the Qing Dynasty, Jingdezhen's export porcelain produced exclusively for the European and American markets was the main representative of China's export porcelain.

纹章瓷盘
A Plate with Coat of Arms

这一时期，中国外销瓷最早的订单，是葡萄牙王室于 15 世纪初在景德镇定制的有其徽号的青花瓷器。后来，欧洲的王室、贵族、公司、城市，都在中国定制绘有其徽号和纹章的瓷器，统称为纹章瓷。

During this period, the earliest order for Chinese export porcelain was blue and white porcelain bearing the family's coat of arms made in Jingdezhen in the early 15th century, which was customized by the Portuguese royal family. Afterwards, the royal families, nobles, companies and cities in Europe customized porcelain decorated with crests or heraldic motifs in China, collectively known as "armorial porcelain."

清乾隆广彩满大人纹碗
A Guangcai Mandarin Bowl, Qianlong Period, Qing Dynasty

清代，还有一种外销瓷，是在景德镇制作好白瓷坯胎，然后运往广州，按照欧洲人的要求，根据他们提供的木样，在广州加工彩绘、开炉烘烧，制成织金彩瓷，其又被称作广彩瓷。

In the Qing Dynasty, there was another kind of porcelain for export. Porcelain biscuits made in Jingdezhen were transported to Guangzhou, where they were painted and fired at the open kiln, turning into gold polychrome glazed porcelain according to the needs of European customers, which was called "Guangcai porcelain."

克拉克瓷碗
A Kraak Bowl

17～18 世纪，景德镇在世界上拥有很高的知名度。

From the 17th to 18th centuries, Jingdezhen had a high reputation in the world.

在欧洲，景德镇瓷器成为皇宫的珍品、上流社会珍爱的奇物、人们身份尊卑的象征。批评者把欧洲上层社会对中国瓷器近于疯狂的喜好和收藏叫作"瓷器热症"。在一些欧洲国家的宫廷里，纷纷建造瓷屋，摆设精美的中国瓷器。法国国王路易十四在凡尔赛宫修建了一座瓷宫，重金收购景德镇生产的青花瓷和彩瓷。由于银圆不足，还大量熔化宫里的银器。

In Europe, Jingdezhen porcelain became a treasure of the imperial palace, a rare thing cherished by the upper-class society, and a symbol of people's dignity and inferiority. Critics called the crazy preference and collection of Chinese porcelain by the European upper class as "porcelain disease." In the courts of some European countries, porcelain houses were built one after another, decorated with exquisite Chinese porcelain. Louis XIV, king of France, built a porcelain palace in Versailles Palace and spent lavishly to purchase blue and white porcelain and colored porcelain made in Jingdezhen. Due to the shortage of silver dollars, a large number of silverwares in the palace were melted.

德国萨克森尼选帝侯兼波兰国王奥古斯都二世曾做过一笔交易：用 600 名萨克森尼龙骑兵换了 151 件中国康熙时期的青花瓷瓶。德国德累斯顿茨温格宫是目前欧洲最大的瓷器专项博物馆，收藏的 42000 多件中国历代瓷器藏品中，大部分来源于奥古斯都二世及其继承人的皇家收藏。

German Friedrich August II, Elector of Saxony and King of Poland, once traded 600 Saxon dragoons for 151 blue-and-white vases of the Kangxi period. Dresden's Zwinger Palace, located in Dresden, Germany, is currently the largest porcelain Museum in Europe with over 42,000 pieces of Chinese porcelain collections, the majority of which come from the royal collection of Augustus II and his successors.

镶嵌贵金属的中国瓷器
Chinese Porcelain Inlaid with Precious Metals

因为大量进口中国瓷器，时间长了，欧洲各个国家的银子大量消耗，引起恐慌。为了阻止中国瓷器的进口，降低欧洲人继续购买中国产品的欲望，破解中国制瓷的秘方，建立自己的瓷器生产成为他们的当务之急。

Significant quantities of Chinese porcelain were being imported into Europe. As time went on, a large amount of silver was consumed in various European countries, causing panic. In order to prevent the import of Chinese porcelain, reduce Europeans' desire to continue to purchase Chinese products and crack the secret recipe of Chinese porcelain making, the establishment of their own porcelain production has become their top priority.

清圣祖爱新觉罗·玄烨之康熙五十一年（1712年）和六十一年（1722年），在景德镇详尽考察7年的法国传教士佩里·昂特雷科莱（中文名殷弘绪），分别写下两封寄往欧洲的长信，详细介绍了景德镇瓷器原材料配方和烧制工艺，首次为欧洲提供了全面准确的中国制瓷信息。

During the period from the 51st (1712) to the 61st year (1722) of the Kangxi reign of Emperor Shengzu (personal name Aisinjuero Xuanye) of the Qing Dynasty, Pere d'Entrecolles (Chinese name Yin Hongxu), a French missionary, who had made a detailed investigation in Jingdezhen for seven years, wrote two long letters detailing the raw material formula and firing process of Jingdezhen

德国梅森瓷厂胆式瓶

A Gall-bladder Vase, Messein Porcelain Factory, German

porcelain, providing Europe comprehensive and accurate information on the manufacture of Chinese porcelain for the first time.

18 世纪初，欧洲仿烧瓷器成功。

At the beginning of the 18th century, Europe was successful in cracking the secret of firing hard porcelain.

从此，欧洲开启了属于自己的制瓷时代。经过欧洲工匠的不断学习和钻研，加上欧洲瓷器生产迅速搭上工业化的快车，在科学技术和自由贸易的辅助下高速发展。中国瓷器长达 1000 多年一枝独秀的地位被打破。

Since then, Europe has opened its own era of porcelain manufacturing. With the continuous efforts of European craftsmen, European porcelain industry developed at a surprising high speed by keeping pace with the industrial revolution, especially with the assistance of science and technology and free trade. Chinese porcelain was no longer in a dominant position lasting for more than 1000 years.

英国青花柳树图案瓷盘
A British Blue and White "Willow Pattern" Plate

欧洲瓷器很快在国际市场上崭露头角，中国瓷器的出口量则出现滑坡。1791 年，英国政府下令停止进口中国瓷器。第二年春天，最后一批以英国东印度公司名义进口的中国瓷器输入英国，从此成批进口中国瓷器的历史宣告结束。

European porcelain soon emerged as a major player in the international market, while the export of Chinese porcelain declined. In 1791, the British government ordered to stop importing Chinese porcelain. With the last batch of Chinese porcelain imported in the name of the British East India Company into the UK next spring, the history of importing porcelain wares in batches from China had come to an end.

中国制瓷业日渐衰败之时，欧洲、日本、美国的制瓷业却扶摇直上，垄断了高端瓷器市场。1840 年鸦片战争爆发，中国外销瓷走到尽头，最终被挤出国际市场。

While China's porcelain industry was declining, European porcelain, Japanese porcelain and American porcelain were soaring, monopolizing the high-end porcelain market. In 1840 when the Opium War broke out, Chinese export porcelain reached its rock bottom and was finally squeezed out of the international market.

鸦片战争后，景德镇制瓷业逐年衰落。

After the Opium War, Jingdezhen porcelain industry declined year by year.

清宣宗爱新觉罗·旻宁之道光年间（1821～1850 年），景德镇有瓷窑270～290 座。到穆宗爱新觉罗·载淳之同治八年（1869 年），只剩下了 60座。太平军与清军多次激战于景德镇，瓷业几近毁尽。洋瓷依仗特权关税与机器生产优势大举倾销，也导致景德镇瓷器销售在国内市场日益缩小。

During the Daoguang reign of Emperor Xuanzong (personal name Aixinjueluo Minning) of Qing Dynasty (1821-1850), there were 270-290 porcelain kilns in Jingdezhen. By the eighth year of the Tongzhi reign (1869) of Emperor Muzong (personal name Aixinjuero Zaichun), there were only 60 kilns left. Because Taiping troops had many fierce battles with the Qing forces in Jingdezhen, the porcelain industry was almost destroyed. Relying on the privileged tariff and the advantages of machine production, foreign porcelain was dumped on a large scale, which also led to the shrinking sales of Jingdezhen porcelain in the domestic market.

第四编

兴于当代

IV Reviving in the Contemporary Era

景德镇市

Jingdezhen City

中华民国元年（1912 年），废府及直州。民国三年（1914 年），江西全省分设四道，浮梁属浔阳道。辛亥革命（1911～1912 年）前，景德镇设有浮梁县丞官职治理，衙门设于御窑厂内。民国五年（1916 年），浮梁县知事陈安呈准省长公署批准，浮梁县治由旧城迁到景德镇，县衙门仍在御窑厂内。民国九年（1920 年），道废，浮梁县直属于省。

In the first year of the Republic of China (1912), the territorial administration organized in superior prefectures (fu) and prefectures directly under the provincial government (zhizhou) were withdrawn. In the third year of the Republic of China (1914), Jiangxi Province was divided into four circuits (dao), and Fuliang belonged to Xunyang circuit (dao). Before the Revolution of 1911, or the Xinhai Revolution (1911-1912), Jingdezhen was placed under the governance of Fuliang county magistrate, and its county government office, also named Yamen in Chinese, was set up in the imperial kiln factory. In the fifth year of the Republic of China (1916), after an application was submitted to the office of the prefecture by Chen An, the magistrate of Fuliang County, Fuliang County Government was approved to move from the old city to Jingdezhen, but the County Yamen was still at the imperial kiln factory. In the ninth year of the Republic of China (1920), as

景德镇老里弄

The Old Lane in Jingdezhen

circuits (dao) were withdrawn, Fuliang County directly belonged to the province.

民国初期，景德镇短暂设市。《景德镇文史资料》记载，最早用景德镇市冠称的机构，始见于以国共合作为基础的中国国民党景德镇市党部。民国十五年（1926 年），江西省政府派张田民为景德镇市第一任市长，于民国十六年（1927 年）1 月到景德镇，筹组景德镇市行政公署，公署设在御窑厂官邸。约在民国十八年（1929 年），景德镇市建制撤销，仍为浮梁县管辖下的一个镇。

At the beginning of the Republic of China, Jingdezhen became a city for a short term. According to *the Records of Jingdezhen Cultural and Historical Materials*, the earliest organization named Jingdezhen city was first found at the Party Headquarter of the Chinese Kuomintang of Jingdezhen City based on the cooperation between the Kuomintang and the Communist Party of China. In 1926, the 15th year of the Republic of China, Jiangxi Provincial Government appointed Zhang Tianmin as the first mayor of Jingdezhen city. In January 1927, the 16th year of the Republic of China, he went to Jingdezhen to prepare for the formation of Jingdezhen Municipal Administrative Office, which was located in the imperial kiln factory. In 1929, about the 18th year of the Republic of China, the organizational system of Jingdezhen city was withdrawn and it was still a town under the jurisdiction of Fuliang County.

民国十八年（1929 年），景德镇陶务局成立。民国十九年（1930 年），景德镇陶务局并入省立工业试验所。民国二十三年（1934 年），江西陶业管理局成立，设在景德镇。

In 1929, the 18th year of the Republic of China, Jingdezhen Ceramic Affairs Bureau was established. In 1930, the 19th year of the Republic of China, Jingdezhen Ceramic Affairs Bureau was incorporated into the Provincial Industrial Experiment Institute. In 1934, the 23rd year of the Republic of China, Jiangxi Ceramics Administration Bureau was established, which was based in Jingdezhen..

民国二十一年（1932 年），江西划分为 13 个行政区，浮梁县属第四行政区。民国二十四年（1935 年），缩减为 8 个行政区，浮梁县属第五行政区。同年，第五行政区督察专员公署由鄱阳迁至浮梁景德镇，署址设在莲花塘。

In 1932, the 21st year of the Republic of China, Jiangxi was divided into 13 administrative regions, and Fuliang County belonged to the fourth administrative one. In 1935, the 24th year of the Republic of China, Jiangxi was reduced to eight administrative regions, and Fuliang County belonged to the fifth one. In the same year, the office of the inspector general of the fifth administrative region was

景德镇莲花塘

Lianhuatang in Jingdezhen

moved from Poyang to Fuliang, Jingdezhen, and its office was located in Lianhuatang.

清末和民国时期，景德镇制瓷业开启向现代制瓷工业的艰难转型。

During the late Qing Dynasty and the Republic of China, Jingdezhen porcelain
industry began its difficult transformation process to modern porcelain industry.

清朝瓦解，民国建立，御窑厂良工四散，禁令废弛，以往所不敢仿制
的贡品，"今则无不敢矣"。御窑垄断格局的打破，使瓷业生产环境宽松，
仿古瓷异常兴旺，市场红火，数量庞大，种类丰富，制瓷技艺普遍提高。

The Republic of China was formed when the Qing Dynasty fell. Skillful
craftsmen of the imperial kiln factory fled in all directions. As the relevant ban re-
leased, artisans began to imitate "tribute porcelains" that they didn't dare to do
before. With the break of the monopoly of the imperial kilns, the manufacture of

民国粉彩婴戏纹花盆
A Famille-rose Flower Pot with Children
Playing, Chinese Republican Period

porcelain tended to be free from restraint. The market of antique porcelain was extremely flourishing and prosperous in a large quantity and rich variety. Moreover, the porcelain handicraft skills were generally improved.

此时，景德镇制瓷业也呈现出变革创新之势，创办了江西瓷业公司，开办了中国陶业学堂。清末民初，景德镇先后从国外引进煤窑窑炉设计及烧成，新法选矿及粉碎，机械练泥、成型及吹釉，石膏的使用及注浆成型，釉上、釉下贴花装饰，氧化钴的使用及新彩装饰等先进制瓷技术。

At that time, Jingdezhen porcelain industry also showed the trend of reform and innovation. Jiangxi Porcelain company, as well as China Ceramics School, was established. Beginning from the end of the Qing Dynasty and the early Republic of China, Jingdezhen successively introduced advanced porcelain manufacturing technology and procedure from abroad, including design and firing of kilns using coal as fuel, new method of mineral processing and crushing, mechanical mud refining, forming and glaze blowing, plaster moulding and slip casting, overglaze and underglaze decal decoration, the application of cobalt oxide and new color decoration, etc.

民国粉彩花木兰从军瓶

A Famille-rose "Hua Mulan Joined the Army" Vase,
Chinese Republican Period

　　艺术瓷创新也呈现新的面貌。清末开始，安徽籍艺人程门、金品卿、王少维、汪晓棠等将黄公望的中国画浅绛彩画法运用于瓷器创作，创立全新的陶瓷绘画艺术形式，以及诸多新样式、新工艺。随后，新粉彩瓷画的兴盛、珠山八友画派的崛起等，对日后景德镇陶瓷艺术的发展产生了重要影响。

　　The innovation of artistic porcelain also presented a new look. Since the late Qing Dynasty, Anhui artists Cheng Men, Jin Pinqing, Wang Shaowei and Wang Xiaotang applied Huang Gongwang's technique of Qianjiangcai (enamels) landscape painting to porcelain creation, creating a new art form of ceramic painting, as well as many new styles and processes. Subsequently, the prosperity of new famille rose painting and the rise of "The Eight Friends of Zhushan" painting school had a significant influence on the later development of Jingdezhen ceramic art.

但是，由于社会动荡、战乱频繁、生产关系落后，加上景德镇瓷业行帮、商会制度约束等多重因素，民国时期景德镇制瓷业的发展依然十分缓慢。

However, due to social unrest, frequent wars, backward production relations plus other multiple factors, such as Jingdezhen porcelain guild, constraints of chamber of commerce system, the development of Jingdezhen porcelain industry was still very slow during the period of the Republic of China.

1949年4月29日上午10时，中国人民解放军第二野战军第五兵团十七军四十九师进入景德镇，浮梁县及景德镇宣告解放。4月30日，景德镇市工商局成立，内设陶瓷科，专管陶瓷工业。

At 10 a.m. on April 29, 1949, as the 49th division of the 17th army of the fifth corps of the second field army of the Chinese People's Liberation Army entered Jingdezhen, Fuliang County and Jingdezhen were liberated. On April 30, Jingdezhen Administration for Industry and Commerce was established within a ceramics section in charge of the ceramic industry.

1949年5月后，景德镇从浮梁县划出，设景德镇市委，成立景德镇人民市政府，属华东局赣东北区党委领导的直辖市，景德镇市委与浮梁地委两块牌子一套领导班子。同年8月，赣东北地区划归江西省建制，区党委撤销。江西省委接管后，设乐平地委，下辖乐平、浮梁等7县和景德镇市。乐平地委初驻乐平县城，同年11月，迁至景德镇市，改称浮梁地委。

After May 1949, with Jingdezhen transformed from Fuliang County, Jingdezhen Municipal Party Committee was established as well as the People's Government of Jingdezhen Municipality, a municipality directly under the central government led by the Party committee of Northeast Jiangxi of East China Bureau. Jingdezhen Municipal Party Committee and Fuliang Prefectural Party Committee were two organizations with a set of leading groups. In August of the same year, as the northeastern area of Jiangxi Province was put under the organizational system of Jiangxi Province, its District Party Committee was withdrawn. After the District Party Committee was taken over by Jiangxi Provincial Party Committee, Leping Prefectural Party Committee was established, which governed seven counties including Leping, Fuliang and so on and Jingdezhen city. Leping Prefectural Party Committee was first stationed in Leping county. In November of the same year, it moved to Jingdezhen city and was renamed as Fuliang Prefectural Party Committee.

浮梁古县衙
Fuliang Ancient Government Office

1952 年 8 月，江西省委决定，撤销浮梁地委，并入上饶地委，驻上饶市办公，景德镇市划出为省辖市。1953 年 6 月 13 日，中南局批复文件正式下达。1960 年，浮梁县撤销，所属各区乡划归景德镇市管辖，设立为鹅湖区和蛟潭区。

In August 1952, Jiangxi Provincial Party Committee decided that the Fuliang Prefectural Party Committee was withdrawn, which was incorporated into the Shangrao Prefectural Party Committee, with its workplace in Shangrao City, and Jingdezhen city was designated as a provincial city. On June 13, 1953, the reply document of Central South Bureau was officially issued. In 1960, Fuliang County was withdrawn, with its districts and townships under the juris diction of Jingdezhen City, which was transformed into Ehu District and Jiaotan District.

1983 年 7 月，上饶地区乐平县、原波阳县（现鄱阳县）鲇鱼山乡和荷塘垦殖场划属景德镇市。1988 年 10 月，浮梁县恢复建制。1992 年 9 月，乐平县撤县建市。至此，形成了景德镇市延续至今的乐平市、浮梁县、珠山区和昌江区行政区划格局。

In July 1983, Leping County, Nianyushan Township and Hetang Reclamation Farm, former Boyang county (now Poyang County) in Shangrao Prefecture were merged into Jingdezhen city. In October 1988, the organizational system of Fuliang County was restored. In September 1992, Leping County was approved to repeal county system and set up city system. So far, the administrative division pattern of Leping City, Fuliang County, Zhushan District and Changjiang District has been formed.

十大瓷厂

The Top Ten Porcelain Factories

中华人民共和国成立之初，景德镇只有私营瓷厂 2540 户，平均每户不足 5 人，规模最大的余鼎顺瓷厂也只有 137 人，最小的仅有雇工 1 人。瓷业生产已奄奄一息。

At the beginning of the founding of the People's Republic of China in 1949, there were only 2,540 private porcelain factories in Jingdezhen, with an average of less than 5 people per household. There were only 137 workers in the largest Yudingshun Porcelain Factory, while the smallest factory employed only one worker. Porcelain production was on its last legs.

景德镇鸟瞰

A Bird's-eye View of Jingdezhen

1950 年 4 月 15 日，景德镇最早的国营瓷厂——江西省景德镇市建国瓷业公司成立，1952 年 10 月 27 日更名为建国瓷厂。1950 年 5 月 4 日，国营景德镇市瓷业公司成立。同年 5 月 14 日，景德镇市瓷业工会成立。

As the earliest state-owned porcelain enterprise in Jingdezhen, Jiangxi Jianguo Porcelain Company was established on April 15, 1950, which was renamed Jianguo Porcelain Factory on October 27, 1952. On May 4,1950, the state-owned Jingdezhen Ceramic Company was established. On May 14, 1950, Labor Union of Jingdezhen Ceramic Industry was established.

颜色釉花瓶
A Colored Glaze Vase

从 1952 年底起，政府开始对陶瓷企业进行社会主义改造。至 1956 年 1 月 18 日，公私合营完成。景德镇相继组建十几家大型国营瓷厂，人们统称为"十大瓷厂"。建国瓷厂之外，又先后建立了人民、新华、艺术、东风、景兴、红星、光明、红旗、宇宙、为民、华风、古窑、景陶、华电等全民所有制瓷厂，红光、曙光、雕塑、卫华、新光（陶彩）等江西省陶瓷工业公司所属集体所有制瓷厂，以及跃进、福利、立新、红卫、昌虹等区办和大厂下属配套集体所有制瓷厂，还有数十家校办和镇办瓷厂。

Since the end of 1952, the government had started to carry out socialist transformation to the ceramic enterprises. By January 18, 1956, the public-private partnership was completed. Jingdezhen has successively established more than a dozen large-scale state-owned porcelain factories, which are collectively referred to as "the top ten porcelain factories." Besides Jianguo Porcelain Factory, other state-owned porcelain enterprises, such as People's Porcelain Factory, Xinhua Porcelain Factory, Art Porcelain Factories, Dongfeng Porcelain Factory, Jingxing Porcelain Factory, Red Star Porcelain Factory, Guangming Porcelain Factory, Red Flag Porcelain Factory, Universe Porcelain Factory, Weimin Porcelain Factory, Huafeng Porcelain Factory, Ancient Kiln Factory, Jingtao Porcelain Factory, Huadian Porcelain Factory, collectively owned porcelain factories affiliated by Jiangxi Ceramic Industry Company, such as Hongguang Porcelain Factory, Shuguang Porcelain Factory, Statuary Porcelain Factory, Weihua Porcelain Factory, Xinguang (Taocai), Porcelain Factory and other collectively owned porcelain factories affiliated by the district or some large factories, including Yuejing, Fuli, Lixin, Hongwei,Changhong, were successively established. In addition, there were dozens of porcelain factories run by school and town.

景德镇制瓷业告别手工作坊，全面进入机械化和标准化。

As handicraft workshops were terminated, Jingdezhen porcelain industry fully adopted mechanization and standardization.

景德镇陶瓷生产方式和生产关系实现重大变革，产业结构也发生历史性变化，由皇家用瓷和工艺瓷生产，发展成日用瓷、工艺美术瓷、建筑瓷、卫生瓷、工业用瓷、电子陶瓷、特种陶瓷、高技术陶瓷等多门类的陶瓷生产体系。

粉彩花瓶

A Famille-rose Vase

While the modes and relations of Jingdezhen ceramic production have undergone a major reform, its industrial structure has also undergone historic changes. The production of imperial porcelain and arts and crafts porcelain has developed into a ceramic production system of various categories, including daily porcelain, arts and crafts porcelain, architectural porcelain, sanitary porcelain, industrial porcelain, electronic ceramics, advanced ceramics, high-tech ceramics and so on.

郭沫若诗云："中华向号瓷之国，瓷业高峰是景都。"

Guo Moruo wrote a poem about Jingdezhen, in which he wrote: "China has long been called the kingdom of porcelain, and the capital of this kingdom is Jingdezhen."

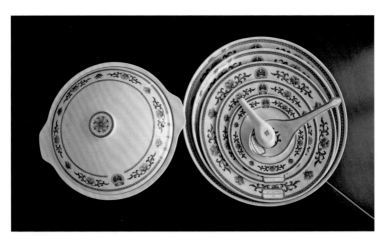

国徽瓷餐具
National Emblem Tableware

十大瓷厂的生产，从品种到产量、从器型到花面，均由行业管理部门计划指定，逐步形成了各个厂的特色产品。光明瓷厂和红光瓷厂生产青花玲珑瓷，人民瓷厂生产青花瓷，艺术瓷厂生产粉彩瓷，建国瓷厂生产颜色釉瓷，雕塑瓷厂生产雕塑美术瓷，曙光瓷厂生产大件缸、钵、箭筒，宇宙瓷厂、为民瓷厂、红星瓷厂以生产釉上彩绘、贴花为主，景兴瓷厂生产传统正德器，东风瓷厂生产壶类等琢器品种，红旗瓷厂生产斗碗饰以缠枝莲青花装饰的品种为主，新华瓷厂生产民族用瓷，华风瓷厂生产饰以青花装饰的宾馆酒店用瓷及配套瓷等。

The production of the top ten porcelain factories, whether from variety to output or from body shape to decorative patterns, is planned and designated by the industry management department, gradually forming the characteristic products of each factory. Guangming Porcelain Factory and Hongguang Porcelain Factory specializes in blue and white porcelain with "rice grain" pattern; People's Porcelain Factory specializes in blue and white porcelain; Arts Porcelain Factory specializes in "famille rose" (Fencai porcelain); Jianguo Porcelain Factory specializes in high temperature colored glazes porcelain; Statuary Porcelain Factory specializes in sculpture art porcelain; Shuguang Porcelain Factory specializes in large-size vats, bowls and quivers; Universal Porcelain Factory, Weimin Porcelain Factory and Red Star Porcelain Factory specializes in products with overglaze color painting and decal decoration; Jingxing Porcelain Factory specializes in traditional Zhengde style wares; Dongfeng porcelain factory specializes in hollow porcelain like pots; Red Flag Porcelain Factory mainly specializes in Dou bowls decorated with blue and white tangled lotus pattern; Xinhua Porcelain Factory specializes in ethnic style household-use porcelain, and Huafeng Porcelain Factory specializes in blue and white porcelain products for hotel and restaurant and accessory products.

由此，十大瓷厂形成分工多样、产能巨大的规模效益。

As a result, the top ten porcelain factories had generated economies of scale with diverse division of labor and huge production capacity.

在发展顺利时期，十大瓷厂年均产瓷 1 亿多件，最高的年份产量达 4

亿件，为国家创外汇 5000 多万美元。日用瓷大批量地来自景德镇，景德镇也成为国宴瓷首选制作地。中华人民共和国成立初期，景德镇瓷器主要出口苏联和东欧。之后，逐步扩大到世界 130 多个国家和地区。20 世纪八九十年代，景德镇瓷业发展达到顶峰，陶瓷产量及销量都远远超过了历史纪录。

During the period of smooth development, the top ten porcelain factories produced more than 100 million pieces per year, with the highest annual output of 270 million pieces, creating foreign exchange of more than 50 million US dollars for the country. Large quantities of household porcelain came from Jingdezhen, and Jingdezhen became the preferred place for producing the state banquet porcelain. During the early period of the People's Republic of China, Jingdezhen porcelain was mainly exported to the Soviet Union and Eastern Europe. Afterwards, it gradually expanded to more than 130 countries and regions in the world. In the 1980s and 1990s, Jingdezhen porcelain industry reached its peak, with the output and sales of ceramics far exceeding the historical record.

青花梧桐餐具
Blue and White "Willow Pattern" Tableware

改革开放以后，景德镇瓷业生产步入新时期，焕发出新活力，发明和创造了色釉彩、综合彩、现代陶艺、现代青花、釉中彩等大量新彩类、新形式、新技法，以及相应的新工艺、新材料。

Since the reform and opening up, Jingdezhen porcelain industry has entered a new period and radiated new vitality, stimulating the invention and creation of a large number new color categories, new forms, new techniques, as well as corresponding new processes and new materials, including colored glaze decoration, comprehensive-colored glaze decoration, modern ceramic art, modern blue and white, in-glaze decoration and so on.

沧海桑田，兴衰不测。

The city has been through many vicissitudes.

在社会转型和经济转轨当中，景德镇十大瓷厂逐渐步入生存危机。从1995 年下半年开始，景德镇所属国营瓷厂先后停产。最后停产的国营瓷厂是宇宙瓷厂，时间为 2002 年 9 月。

In the social transformation and economic transition, the top ten porcelain factories in Jingdezhen have gradually stepped into the survival crisis. Since the second half of 1995, Jingdezhen's state-owned porcelain factory has successively stopped production. The last state-owned porcelain factory to stop production was Universal Porcelain Factory in September 2002.

瓷业复兴

The Rival of Porcelain Industry

景德镇制瓷业再次进入转型期。

Jingdezhen porcelain industry has entered the transition period again.

2005 年，乐天陶社选定原雕塑瓷厂作为文化创意产业园，成为景德镇陶瓷文创产品制作、运营、销售模式的发端，形成第一批工业遗产内的文化创意产业集群，带动景德镇陶瓷大学老校区陶瓷文化创意街的兴起。2010 年，景德镇打造了依托原艺术瓷厂的红店街。

三宝瓷谷远眺
Overlooking Sanbao International Ceramic Valley

In 2005, the Pottery Workshop selected the original sculpture porcelain factory as the cultural and creative industrial park, which has become the originator of the production, operation and sales mode of Jingdezhen ceramic cultural and creative products, forming the first batch of cultural and creative industry clusters in the industrial heritage, and driving the rise of ceramic cultural and creative street in the old campus of Jingdezhen Ceramic University. In 2010, Jingdezhen built the Hongdian Street at the original site of Art Porcelain Factory.

初建中的陶溪川创意街区
Taoxichuan Ceramic Art Avenue under Construction

2011年，景德镇在发展性保护近代工业遗存的理念指导下，结合北京798、上海M50等老工厂改文创区的发展思路，逐渐开始对十大瓷厂工业遗存进行整合，斥巨资打造了依托原国营宇宙瓷厂的陶溪川国际陶瓷文化创意园。

In 2011, under the guidance of the concept of developmental protection of modern industrial relics, Jingdezhen, through combining the development idea of transforming old factories such as Beijing 798 and Shanghai M50 into cultural and creative zones, gradually began to integrate the industrial relics of the top ten

porcelain factories, investing a large sum of money to build Taoxichuan International Ceramic Cultural and Creative Industrial Park at the original factory site of the former state-owned Universal Porcelain Factory.

从 2010 年到 2015 年，景德镇陶瓷文化创意产业数量逐渐增加，分布逐渐由老城区向外围扩散。此时，东到景德镇陶瓷大学湘湖校区，西至陶瓷文化创意新区，南到三宝国际瓷谷，北至陶瓷工业园区，景德镇陶瓷产业及文创产业整体呈现不规则的"带状组团 + 指状"形态。

From 2010 to 2015, the number of Jingdezhen ceramic cultural and creative industries gradually increased, and the distribution slowly spread from the old urban area to the urban periphery. By now, from Xianghu campus of Jingdezhen Ceramic University in the East to the ceramic cultural and creative new area in the west, from Sanbao International Ceramic Valley in the south to the ceramic industrial park in the north, Jingdezhen ceramic industry and cultural and creative industry as a whole emerge an irregular form of "Banded Group plus Finger Shape."

景德镇的民间创意产业实体也开始出现了跨越式的发展。到 2015 年底，景德镇已经拥有 15 家文化创意产业基地，以及超过 5000 家的创意产业经济实体，陶瓷创意产业集群效应开始显现。

The folk creative industry entities in Jingdezhen have also been developed by leaps and bounds. At the end of 2015, Jingdezhen had 15 cultural and creative industry bases and more than 5,000 creative industry economic entities, and the cluster effect of ceramic creative industry began to appear.

因瓷而生的景德镇，如今因瓷而变。

Jingdezhen, born of porcelain, is now changed by porcelain.

景德镇正在打造跨界混合业态。信息化的发展，新材料、新工艺的突破，给陶瓷创意创新提供了广阔的空间和舞台。景德镇不断加强基础研究、完善公共服务平台、优化创意创新环境，积极推动陶瓷艺术的融合创新，形成了以陶瓷为核心的跨界混合业态。

Jingdezhen is building a cross-border mixed business. The development of information technology and the breakthrough of new materials and processes provide a broad space and stage for ceramic creative innovation. Jingdezhen has formed a cross-border mixed business form with ceramics as the core through continuously strengthening basic research, improving the public service platform, optimizing the creative innovation environment, actively promoting the integration and innovation of ceramic art.

建国陶瓷文化创意园
Jianguo Ceramic Cultural and Creative Park

景德镇构筑大陶瓷发展格局，从日用陶瓷到精细陶瓷、特种陶瓷、科技陶瓷等，景德镇正在打造陶瓷创新链、文化链、产业链。无论是日用、艺术、创意、建筑等陶瓷，还是高科技陶瓷，门类齐全、工艺丰富，均占据国内制高点。高技术陶瓷发展迅猛，由改革开放初期的几乎空白发展到目前 60.9 亿元规模，是各类陶瓷产品中增速最快的。同样快速发展的还有创意陶瓷，2017 年景德镇艺术陈设瓷产值达 126.2 亿元。

The development pattern of "Great Ceramics", ranging from daily-use ceramics to fine ceramics, special ceramics, scientific and technological ceramics, is constructed in Jingdezhen, with ceramic innovative chain, cultural chain and industrial chain being built. whether ceramics for daily use, art, creativity, architecture but also high-tech ceramics are marked by complete categories and rich processes, which occupy the commanding heights in China. High tech ceramics are developing rapidly, from almost blank in the early stage of reform and opening up to the current scale of 6.09 billion yuan, showing the fastest growth among all kinds of ceramic products. Creative ceramics are also developing rapidly. In 2017, the output value of Jingdezhen art display porcelain reached 12.62 billion yuan.

2019 年，景德镇陶瓷产业产值达 523 亿元，艺术陈设瓷、高技术陶瓷占陶瓷总产值的半壁江山。从传统制瓷到现代文旅，从创意空间到工业服务平台，景德镇正不断拓展陶瓷的外延。

In 2019, the output value of Jingdezhen ceramic industry reached 52.3 billion yuan with artistic display ceramics and high-tech ceramics accounting for half of the total ceramics output. From traditional porcelain handcraft to contemporary culture and tourism industry, from creative space to industrial service platform, Jingdezhen is constantly expanding the extension of ceramics.

2019 年 8 月，国家发改委和文旅部印发《景德镇国家陶瓷文化传承创新试验区实施方案》，提出建设国家陶瓷文化保护传承创新基地、世界著名

景德镇御窑博物馆展厅

Exhibition Hall of Jingdezhen Imperial Kiln Museum

陶瓷文化旅游目的地和国际陶瓷文化交流合作交易中心的发展目标，推动景德镇成为集中展示中华陶瓷文化的瓷都、全国乃至世界的陶瓷产业标准和创新中心，把景德镇打造成世界一流的国际文化旅游名城，把试验区建设成为促进全球文明互鉴的重要桥梁和高端陶瓷文化贸易出口区。

In August 2019, the National Development and Reform Commission (NDRC) and the Ministry of Culture and Tourism issued "The Implementation Plan of Jingdezhen National Ceramic Culture Inheritance and Innovation Pilot Zone", which proposed the development goal of building a national ceramic culture protection, inheritance and innovation base, a world-famous ceramic culture tourism destination and an international ceramic culture exchange and cooperation center. The programme aims to promote Jingdezhen to become a ceramic capital displaying Chinese ceramic culture, a domestic and even international ceramic industry standards and innovation center, a world-class international cultural tourism city. According to the plan, the zone, gradually but steadily, will be built into an important bridge to promote mutual learning of global civilizations and an export area of high-end ceramic culture trade.

兴衰可轮回，发展无穷期。

Rise and fall can be reincarnated, but the future development is infinite.

世界上可能没有一种产业和一种产品，像瓷器一样长盛不衰，充满无限的需求和发展前景，居于朝阳产业之列。瓷运连着国运，千年窑火见证历史变迁。千年窑火不熄，景德镇走上了一条转型发展的新路。

景德镇昌江夜景
Night View of the Chang River

So far, perhaps there is no industry or product in the world, which is as prosperous as porcelain, full of unlimited demand and development prospects, ranking among the sunrise industries. The porcelain progress is connected with the national progress; therefore, the kiln fire has witnessed the historical changes of more than 1,000 years. Millennium kiln fire never goes out. Jingdezhen has embarked on a new road of transformation and development.

圣火城雕
City Sculpture "Sacred fire"

参考书目
Bibliography

魏望来.景德镇文化研究（第一、二、三辑）[M].北京：中国文史出版社 2017 年、2018 年版

Wei Wanglai. *Jingdezhen Culture Studies* (Series I, II and III) [M]. Beijing: China Culture and History Press, 2017, 2018.

魏望来.景德镇文化概论[M].武汉：武汉大学出版社 2019 年版

Wei Wanglai. *Introduction to Jingdezhen Culture* [M]. Wuhan: Wuhan University Press, 2019

魏望来.景德镇陶瓷文化读本——数字里的故事[M].南昌：江西高校出版社 2019 年版

Wei Wanglai. *Insights into Jingdezhen Ceramic Culture-Stories in Numbers* [M]. Nanchang: Jiangxi University Press, 2019.

艾春龙，魏望来.遇见景德镇：老城旧事[M].广州：广东人民出版社 2019 年版

Ai Chunlong and Wei Wanglai. *Encounter Jingdezhen: The Story of Old*

Town [M]. Guangzhou: Guangdong People's Publishing House, 2019.

周銮书.景德镇史话[M].南昌：江西人民出版社 2004 年版
Zhou Luanshu. *History of Jingdezhen*[M]. Nanchang: Jiangxi People's Publishing House, 2004.

林景梧.瓷都史话[M].南昌：百花洲出版社 2004 年版
Lin Jingwu. *History of Porcelain Capital*[M]. Nanchang: Baihuazhou Literature and Art Publishing House, 2004.

景德镇市地方志编纂委员会.景德镇市志（第一卷）[M].北京：中国文史出版社 1991 年版
Jingdezhen Local Chronicles Compilation Committee. *Jingdezhen City Chronicle (Volume I)* [M]. Beijing: China Culture and History Press, 1991.

梁淼泰.明清景德镇城市经济研究[M].南昌：江西人民出版社 2004 年版
Liang Miaotai. *Studies on City Economy of Jingdezhen in Ming and Qing Dynasties*[M]. Nanchang: Jiangxi People's Publishing House, 2004

方李莉.飘逝的古镇：瓷都旧事[M].北京：群言出版社 2001 年版
Fang Lili, *Elapsing Ancient Town—Memories of the Ceramic City*[M]. Beijing: Qunyan Publishing House, 2001.

钟健华，陈雨前.景德镇陶瓷史[M].南昌：江西人民出版社 2016 年版
Zhong Jianhua, Chen Yuqian. *History of Jingdezhen Ceramics* [M]. Nanchang: Jiangxi People's Publishing House, 2016.

中国国家人文地理编委会.景德镇[M].北京：中国地图出版社 2016 年版

China National Human Geography Editorial Committee. *Jingdezhen* [M]. Beijing: China Map Publishing House, 2016.

景德镇市政协文史和学习委员会.景德镇文史资料（1～14 辑）[M].南昌：江西高校出版社 2018 年版

The Historical Data and Study Committee of the CPPCC Jingdezhen Municipal Committee. *Jingdezhen Historical Data（bound edition）（1-14 Series）*[M]. Nanchang: Jiangxi Higher Education Press, 2018.

冯先铭.中国陶瓷[M].上海：上海古籍出版社 2001 年版

Feng Xianming. *Chinese Ceramics* [M]. Shanghai:Shanghai Ancient Books Publishing House, 2001.

田自秉.中国工艺美术史[M].上海：东方出版中心 2010 年版

Tian Zibing. *History of Chinese Arts and Crafts*[M]. Oriental publishing center, 2010.

罗学正.陶林撷翠：中国古陶瓷史话百题[M].台北：五行图书出版有限公司 2004 年版

Luo Xuezheng. *The Stories of Chinese Ancient Ceramics*[M]. Taibei:Taiwan Wuxing Book Publishing Co., Ltd., 2004.

阎崇年.御窑千年[M].北京：生活·读书·新知三联书店 2017 年版

Yan Chongnian.*Thousand Years of Imperial Kilns* [M]. Beijing: SDX Joint Publishing Company, 2017.

郑云云.千年窑火[M].南昌：江西人民出版社 2007 年版

Zheng Yunyun. *Millennium Kiln Fire*[M]. Nanchang: Jiangxi People's Publishing House, 2007.

后 记

Postscript

　　《景德镇瓷业发展简史》是由景德镇学院、景德镇国际丝路学院、景德镇发展研究院出品，景德镇学院党委副书记、景德镇发展研究院院长、景德镇国际丝路学院执行院长王丽心，景德镇陶瓷文化研学研究院院长、景德镇发展研究院常务副院长兼秘书长、高级编辑魏望来，景德镇学院外国语学院院长、教授朱练平著译，主要用于开展景德镇陶瓷文化研修和研学的辅助读物。

A Brief History of Jingdezhen Porcelain Industry is produced by Jingdezhen University, Jingdezhen International Silk Road Academy and Jingdezhen Development Research Institute. It was written by Wang Lixin, deputy secretary of the Party Committee of Jingdezhen University, president of Jingdezhen Development Research Institute and executive president of Jingdezhen International Silk Road Academy, and Wei Wanglai, senior editor, president of Jingdezhen Ceramic Culture Research Institute, executive vice president and secretary general of Jingdezhen Development Research Institute, and translated by Professor Zhu Lianping, dean of the School of Foreign Languages of jingdezhen University, which is mainly served as an supplementary reading for the study of Jingdezhen ceramic culture.

　　研究和阐释景德镇陶瓷文化的书籍，虽不能说是汗牛充栋、浩如烟海，但亦可谓林林总总、蔚然可观。为体现研修和研学的普适性、简约性、通俗性，本书在参考和采信大量相关文献的基础上，删繁就简，力求用提纲挈领式的语句，勾勒出清晰可见的景德镇瓷业基本脉络和发展轨迹。为此，本书注重了时空的有序交织。一方面，纵向概括为起于汉唐、成于宋元、盛于明清、兴于当代等四个部分，以显示时间上的连续性。另一方面，每部分又横向编入与时间框架相对应的水土宜陶、始于汉世、制器进御、景德置镇、青白瓷系、浮梁瓷局、青花瓷器、多彩时代、天工开物、明御器厂、清御窑厂、器走天下、景德镇市、十大瓷厂、瓷业复兴等重大事件或标志性内容，以呈现空间上的代表性。

　　Although the books on the study and interpretation of Jingdezhen ceramic culture cannot be said to be voluminous and vast, they are also numerous and impressive. Based on a large number of relevant references, this book tries to outline the basic context and development track of Jingdezhen porcelain industry with concise sentences in order to embody the universality, simplicity and popularity of study and training. Therefore, this book pays attention to the orderly interweaving of time and space. On the one hand, the book is vertically divided into four parts, namely, "Starting in the Han and Tang Dynasties", "Maturing in the Song and Yuan Dynasties", "Attaining Its Acme in the Ming and Qing Dynasties" and "Reviving in the Contemporary Era", so as to show the temporal continuity. On the other hand, major events or symbolic contents corresponding to the time frame, including "A Promised Land for Ceramic Manufacturing", "Porcelain Manufacturing Traced Back to the Han Dynasty", "Porcelain as Tribute to the Court", "the

Establishment of Jingde Town", "Two Kiln Systems of Celadon and Green-ish-White Porcelain","Fuliang Porcelain Bureau", "The Colorful Era", "Exploita-tion of the Works of Nature　(Tiangong Kaiwu)","The Imperial Ware Factory of the Ming Dynasties", "The Imperial Kiln Factory of the Qing Dynasty", "Sending Porcelain to All Parts of the World", "Jingdezhen City", "The Top Ten Porcelain Factories", and　"The Rival of Porcelain Industry", are horizontally incorporated in each part to present the representativeness of space.

　　本书主要图片由余银泉、周国华、徐玉光、祝松星、张文江、刘火金、张瑞麟、江景新等提供，部分图片取自中国文史出版社出版的《景德镇文化研究》（第一、二、三辑）。封面图片《窑火》由余银泉先生提供。陈猛先生对本书进行了认真校阅，郑伊娜女士提供了出版服务。在此，一并致以诚挚的谢意。

The majority of photos in this book are provided by Yu Yinquan, Zhou Guo-hua, Xu Yuguang, Zhu Songxing, Zhang Wenjiang, Liu Huojin, Zhang Ruilin, Jiang Jingxin, etc., and some are taken from *Jingdezhen Culture Studies* (Series I, II and III) published by China Culture and History Press. The cover photo "Kiln Fire" is provided by Yu Yinquan.　Mr. Chen Meng carefully reviewed the book and Ms. Zheng Yina provided publishing services. Here, the authors would like to take this opportunity to extend our sincere gratitude for their help and assistance.

　　用区区数万字的篇幅，展现千年瓷都景德镇的风姿，是传承创新陶瓷文化的一次尝试，我们也期盼以此举向历史和工匠表达敬意。限于个人的学识与学养，书中难免会有诸多概述不准、阐释不当之处，敬请方家指正。

In an attempt to inherit and innovate ceramic culture by presenting the brief history of Jingdezhen-the millennium porcelain capital-with limited tens of thousands of words, we also expect to pay tribute to history and craftsmen. Limited to personal knowledge and academic attainment, there will inevitably be some inaccuracies in the overview and improper interpretation in the book. Any comments and suggestions will be appreciated.

著译者

The authors and the translator

2022 年 5 月

May 2022

图书在版编目（CIP）数据

景德镇瓷业发展简史 / 魏望来, 王丽心著 ; 朱练平
译. -- 北京 : 中国文史出版社, 2022.11
  ISBN 978-7-5205-3772-8

  Ⅰ.①景… Ⅱ.①魏… ②王… ③朱… Ⅲ.①陶瓷工
业—手工业史—景德镇市 Ⅳ.①TQ174-092

  中国版本图书馆 CIP 数据核字（2022）第181393号

责任编辑：王文运　　　　　装帧设计：尚俊文化

出版发行：中国文史出版社
社　　址：北京市海淀区西八里庄路 69 号　　邮　编：100142
电　　话：010-81136606　81136602　81136603（发行部）
传　　真：010-81136655
印　　装：北京地大彩印有限公司
经　　销：全国新华书店
开　　本：710mm×1000mm　1/16
字　　数：150 千字
印　　张：10.75
版　　次：2023 年 3 月北京第 1 版
印　　次：2023 年 3 月第 1 次印刷
定　　价：68.00 元